T0250499

BLOCKCHAIN TECHNOLOGY AND THE INTERNET OF THINGS

Challenges and Applications in Bitcoin and Security

BLOCKCHAIN TECHNOLOGY AND THE INTERNET OF THINGS

Challenges and Applications in Bitcoin and Security

Edited by

Rashmi Agrawal
Jyotir Moy Chatterjee
Abhishek Kumar
Pramod Singh Rathore

First edition published 2021

Apple Academic Press Inc.
1265 Goldenrod Circle, NE,
Palm Bay, FL 32905 USA
4164 Lakeshore Road, Burlington,
ON, L7L 1A4 Canada

CRC Press
6000 Broken Sound Parkway NW,
Suite 300, Boca Raton, FL 33487-2742 USA
2 Park Square, Milton Park,
Abingdon, Oxon, OX14 4RN UK

Library and Archives Canada Cataloguing in Publication

Title: Blockchain technology and the Internet of things : challenges and applications in bitcoin and security / edited by Rashmi Agrawal, Jyotir Moy Chatterjee, Abhishek Kumar, Pramod Singh Rathore.

Names: Agrawal, Rashmi, 1978- editor. | Chatterjee, Jyotir Moy, editor. | Kumar, Abhishek, 1989- editor. | Rathore, Pramod Singh, 1988- editor.

Description: Includes bibliographical references and index.

Identifiers: Canadiana (print) 20200270877 | Canadiana (ebook) 20200271237 | ISBN 9781771888974 (hardcover) | ISBN 9781003022688 (ebook)

Subjects: LCSH: Blockchains (Databases) | LCSH: Internet of things. | LCSH: Bitcoin.

Classification: LCC QA76.9.B56 B56 2021 | DDC 005.75/8—dc23

Library of Congress Cataloging-in-Publication Data

Names: Agrawal, Rashmi, 1978- editor. | Chatterjee, Jyotir Moy, editor. | Kumar, Abhishek, 1989- editor. | Rathore, Pramod Singh, 1988- editor.

Title: Blockchain technology and the internet of things : challenges and applications in bitcoin and security / edited by Rashmi Agrawal, Jyotir Moy Chatterjee, Abhishek Kumar, Pramod Singh Rathore.

Description: Burlington, ON, Vanada ; Palm Bay, Florida, USA : Apple Academic Press, 2020. | Includes bibliographical references and index. | Summary: "Blockchain Technology and the Internet of Things looks at the electrifying world of blockchain technology and how it has been revolutionizing the Internet of Things (IoT) and cyber-physical systems (CPS). Aimed primarily at business users and developers who are considering blockchain-based projects, the volume provides a comprehensive introduction to the theoretical and practical aspects of blockchain technology. It presents a selection of chapters on topics that cover new information on blockchains and bitcoin security, IoT security threats and attacks, privacy issues, fault-tolerance mechanisms, and more. Some major software packages are discussed. It also addresses the legal issues currently affecting the field. The information presented here is relevant to current and future problems relating to blockchain technology and will provide the tools to to build efficient decentralized applications. Blockchain technology and the IoT can profoundly change how the world-and businesses-work, and this book provides a window into the current world of blockchains. No longer limited to just Bitcoin, blockchain technology has spread into many sectors and into a significant number of different technologies. This is a book for those who are interested in blockchains but are overwhelmed by the sudden explosion of options"-- Provided by publisher.

Identifiers: LCCN 2020025951 (print) | LCCN 2020025952 (ebook) | ISBN 9781771888974 (hardcover) | ISBN 9781003022688 (ebook)

Subjects: LCSH: Blockchains (Databases)--Industrial applications. | Internet of things--Security measures. | Bitcoin.

Classification: LCC QA76.9.B56 B564 2020 (print) | LCC QA76.9.B56 (ebook) | DDC 005.8--dc23

LC record available at https://lccn.loc.gov/2020025951

LC ebook record available at https://lccn.loc.gov/2020025952

ISBN: 978-1-77188-897-4 (hbk)
ISBN: 978-1-00302-268-8 (ebk)

About the Editors

Rashmi Agrawal, PhD, is presently working as professor at the Manav Rachna International Institute of Research and Studies, Faridabad, India. She earned UGC-NET qualified with 18+ years of experience in teaching and research. She has authored/coauthored more than 40 research papers in various peer-reviewed national/international journals and conferences. She has also edited/authored books with national/international publishers (IGI Global, Apple Academic Press, and CRC Press) and has contributed many chapters in edited books published by IGI global, Springer, Elsevier, and CRC Press. She has also published two patents in renewable energy. Currently she is guiding PhD scholars in sentiment analysis, educational data mining, Internet of Things, brain-computer interface, web service architecture, and natural language processing. She is associated with various professional bodies in different capacities and is a life member of the Computer Society of India, IETA, ACM CSTA, and a senior member of the Science and Engineering Institute (SCIEI).

Jyotir Moy Chatterjee, MTech, is currently working as an Assistant Professor in the IT Department at Lord Buddha Education Foundation (Asia Pacific University of Technology & Innovation), Kathmandu, Nepal. Prior to this he worked as an Assistant Professor at the CSE Department at GD Rungta College of Engineering & Technology (CSVTU), Bhilai, India. He completed MTech in computer science engineering from the Kalinga Institute of Industrial Technology, Bhubaneswar, Odisha, India, and BTech in computer science and engineering from the Dr. MGR Educational & Research Institute, Chennai, Tamil Nadu, India. He has more than 36 international publications, two authored books, two edited books, and five book chapters into his account. His research interests include the cloud computing, big data, privacy preservation, data mining, Internet of Things, machine learning, and blockchain technology. He is a member of various professional societies and attends international conferences.

Abhishek Kumar, PhD, a senior member of IEEE, is currently working as an Assistant Professor in the Research Department at the Chitkara University Institute of Engineering and Technology, Chitkara University, Punjab, India. He has an MTech in Computer Science and Engineering from Government Engineering College Ajmer, Rajasthan Technical University, Kota, India. He has more than eight years of academic teaching experience and has more than 70 publications in reputed peer-reviewed national and international journals and conferences as well as book chapters and 23 edited and authored books. His research area includes artificial intelligence, image processing, computer vision, data mining and machine learning. he has been on the international conference committees of many international conferences. He has been the reviewer or editor of various peer-reviewed journals indexed in SCI and Scopus databases

Pramod Singh Rathore, MTech, is Assistant Professor of Computer Science at Aryabhatt Engineering College and Research Centre, Rajasthan, India, and also a visiting faculty at Government University MDS Ajmer, India His research areas include NS2, computer networks, mining, and database management systems.

Contents

Contributors

Rashmi Agrawal
MRIIRS, Faridabad, Haryana, India

Md. Shajid Ansari
Department of CSE, RSR-RCET, Bhilai, India

Kavita Arora
Manav Rachna International Institute of Research and Studies, Faridabad 121001, Haryana, India

B. Balamurugan
Department of CSE, Galgotias University, Greater Noida, UP, India

Anuja Bansal
Dr. Ambedkar College, Tonk, Rajasthan, India

Shubham Bhardwaj
Netaji Subhas University of Technology, New Delhi, India

Jyotir Moy Chatterjee
Department of IT, LBEF(APUTI), Kathmandu, Nepal

Abha Choubey
Department of CSE, SSGI, Bhilai, India

Siddharth Choubey
Department of CSE, SSGI, Bhilai, India

Kapil Chouhan
ACERC, Ajmer, Rajasthan, India

Somesh Kumar Dewangan
Department of CSE, GD-RCET, Bhilai, India

Pooja Dixit
Dezyne Ecole College, Ajmer, India

Sumathy Eswaran
Dr. MGR Educational and Research Institute, Chennai, Tamil Nadu, India

Vikas Garg
Amity University, Uttar Pradesh, India

Simran Kaur Jolly
MRIIRS, Faridabad, Haryana, India

Tejas Khajanchee
College of Engineering Pune (COEP), Pune, India

Neelu Khare
School of Information Technology and Engineering, VIT University, Vellore, Tamil Nadu, India

Anita Khosla
MRIIRS, Faridabad, Haryana, India

Deepak Kshirsagar
Department of Computer Engineering and Information Technology,
College of Engineering Pune (COEP), Pune, India

P. Srinivas Kumar
Department of CSE, SSUTMS, MP, India

Abhishek Kumar
Computer Science & Engineering Department, Chitkara University Institute of Engineering and
Technology, Chitkara University, Himachal Pradesh, India

Aditi Kumar
Sri Venkateswara College, New Delhi, India

Palvadi Srinivas Kumar
Research Scholar, Department of CSE, SSUTMS, Sehore, MP, India

Hoang Viet Long
People's Police University of Technology and Logistics, Bac Ninh, Vietnam

Geeta Nijhawan
MRIIRS, Faridabad, Haryana, India

Siddharth Sagar Nijhawan
Netaji Subhas University of Technology, New Delhi, India

Jyotiprakash Patra
Department of CSE, SSIPMT, Raipur, India

Manju Payal
Academic Hub, Ajmer, India

D. Preethi
School of Information Technology and Engineering, VIT University, Vellore, Tamil Nadu, India

S. P. Rajagopalan
Dr. MGR Educational and Research Institute, Chennai, Tamil Nadu, India

Pramod Singh Rathore
ACERC, Ajmer, Rajasthan, India

Thota Siva Ratna Sai
Research Scholar, Department of CSE, SSUTMS, Sehore, MP, India

Gurinder Singh
Amity University, Uttar Pradesh, India

Pooja Tiwari
ABES Engineering College, Ghaziabad, Uttar Pradesh, India

B. K. Tripathy
School of Information Technology and Engineering, VIT University, Vellore, Tamil Nadu, India

Tong Anh Tuan
People's Police University of Technology and Logistics, Bac Ninh, Vietnam

Shruti Vashist
Manav Rachna University, Faridabad, Haryana, India

Abbreviations

ABE	attribute-based encryption
ADEPT	autonomous decentralized peer-to-peer telemetry
AI	artificial intelligence
AML	antimoney laundering
AR	augmented reality
BC	blockchain
BCT	blockchain technology
BD	big data
BGP	border gateway protocol
CoAP	constrained application protocol
CIoT	Consumer Internet of Things
CPU	central processing unit
DAC	distributed autonomous corporations
DAO	decentralized autonomous corporation
DDoS	distributed denial of service
DLT	distributed ledger technology
DLT	distributed ledger technology
DNS	domain name system
DPoS	delegated proof of stake
ECDSA	elliptic curve digital signature algorithm
ECU	electronic control units
GHOST	greedy heaviest-observed sub-tree
FPoW	full proofs-of-work
HER	electronic health record
GHOST	greedy heaviest-observed sub-tree
ICMP	internet control messaging protocol
ICS	indicator-centric schema
ICT	information and communication
IoHT	Internet of Healthcare Things
IIoT	Industrial Internet of Things
IoT	Internet of things
KYC	know your customer
LAE	learning as earning

LPWAN	low power wide area network
LSB	lightweight scalable blockchain
M2M	machine-to-machine
MPC	multiparty computing
NAT	network address translation
NG	next generation
OSI	open system interconnection
PBFT	practical Byzantine fault tolerance
P2PKH	pay-to-pubKeyHash
PKI	public key infrastructure
PO	postal operator
PoS	proof of stake
PoW	proof-of-work
PPoW	partial proofs-of-work
RFID	radio frequency identification
RPL	routing protocol
RPM	remote patient monitoring
SHA	secure hash algorithm
SPOF	single point of failure
TPS	transactions per second
UDP	user datagram protocol
V2V	vehicle-to-vehicle
V2I	vehicle-to-infrastructure
WRSU	wireless remote software update

Introduction to the Book

A blockchain is a public ledger of information collected through a network that sits on top of the internet. It is how this information is recorded that gives blockchain its groundbreaking potential. Blockchain technology is not a company, nor is it an app, but rather an entirely new way of documenting data on the internet. The technology can be used to develop blockchain applications, such as social networks, messengers, games, exchanges, storage platforms, voting systems, prediction markets, online shops and much more. In this sense, it is similar to the internet, which is why some have dubbed it "The Internet 3.0."

The information recorded on a blockchain can take on any form, whether it is denoting a transfer of money, ownership, a transaction, someone's identity, an agreement between two parties, or even how much electricity a lightbulb has used. However, to do so requires a confirmation from several devices, such as computers, on the network. Once an agreement, otherwise known as a consensus, is reached between these devices to store something on a blockchain it is unquestionably there, it cannot be disputed, removed or altered, without the knowledge and permission of those who made that record, as well as the wider community. Rather than keeping information in one central point, as is done by traditional recording methods, multiple copies of the same data are stored in different locations and on different devices on the network, such as computers or printers. This is known as a peer-to-peer (P2P) network. This means that even if one point of storage is damaged or lost, multiple copies remain safe and secure elsewhere. Similarly, if one piece of information is changed without the agreement of the rightful owners, there are countless other examples in existence, where the information is true, making the false record obsolete. Blockchain owes its name to how it works and the manner in which it stores data, namely that the information is packaged into blocks, which link to form a chain with other blocks of similar information.

It is this act of linking blocks into a chain that makes the information stored on a blockchain so trustworthy. Once the data is recorded in a block it cannot be altered without having to change every block that came after it, making it impossible to do so without it being seen by the other participants on the network. Blockchain is a type of distributed ledger for maintaining a permanent and tamper-proof record of transactional data. A blockchain functions as a decentralized database that is managed by computers belonging to a peer-to-peer (P2P) network. Each of the computers in the distributed network maintains a copy of the ledger to prevent a single point of failure (SPOF) and all copies are updated and validated simultaneously.

This book is primarily aimed at research peoples who want an introduction to the technology, how it works, the major types of blockchains out there, and some help getting started by using it is your organization. It's been said that blockchain will do for transactions what the Internet did for information. What that means is that blockchain allows increased trust and efficiency in the exchange of almost anything. Blockchain can profoundly change how the world works. If you've ever bought a house, you've probably had to sign a huge stack of papers from a variety of different stakeholders to make that transaction happen. If you've ever registered a vehicle, you likely understand how painful that process can be. You won't even get started on how challenging it can be to track your medical records. Blockchain—most simply defined as a shared, immutable ledger—has the potential to be the technology that redefines those processes and many others. To be clear, when someone talks about blockchain, he isn't talking about Bitcoin. Rather talking about the underlying digital foundation that supports applications *such as* Bitcoin. But the reaches of blockchain extend far beyond Bitcoin. We have tried to equip readers with an understanding of what blockchain is, how it works, and how it can enhance one's business and the industry in which it operates. We also tried to show how the digital future can be with the blockchain technology. Readers learn the fundamentals of blockchain and how this technology revolutionizes transactions and business networks. Readers can also discover the important difference between "blockchain" and "blockchain for business" and what makes blockchain an ideal solution for streamlining business networks. It will provide better understanding of how digital currency process information using complex networks. This book will show readers about the links between Blockchain, Information Technology, Cryptography, and Computer Science applications in a broad perspective. Finally, readers will find out everything they need to spin up a blockchain network today.

Preface

Blockchain Technology and the Internet of Things is designed to introduce newbies to the world of blockchain in a novice fashion. It takes readers through the electrifying world of blockchain technology, and it is aimed at those who want to polish their existing knowledge and understanding of the various concepts of the blockchain system.

This book will act as a guide that teaches how to apply principles and ideas that makes one's life and business better. Readers only need a curious mind to get started with blockchain technology. Once they have grasped the basics, they can easily understand the underlying technology or methodology it follows. It will follow by exploring different types of blockchain with easy-to-follow methods.

In addition to this, you will learn how blockchain has been revolutionizing the Internet of Things (IoT), cyber physical systems (CPS), and how it can affect business processes. By the end of this book, you will not only have solved current and future problems relating to blockchain technology but also be able to build efficient decentralized applications.

This is a book for those who are interested in blockchain but are overwhelmed by the sudden explosion of options. No longer limited to just Bitcoin, blockchain technology has spread into many sectors and a significant number of different technologies.

This book is also aimed primarily at business users and developers who are considering a blockchain-based projects. This book will help orient you to the current world of blockchain, and it introduces the major software projects and packages and covers some of the legal background currently affecting the field.

This book has one goal: to provide a comprehensive introduction to the theoretical and practical aspects of blockchain technology. This book contains all the material that is required to fully understand blockchain technology. After reading this book, readers will be able to develop a deep understanding of the inner workings of blockchain technology and will be able to develop blockchain applications. This book covers all the topics relevant to blockchain technology, including cryptography, cryptocurrencies, and various other platforms and tools used for block-chain development.

— **Rashmi Agrawal**
Jyotir Moy Chatterjee
Abhishek Kumar
Pramod Singh Rathore

Acknowledgment

I would like to acknowledge the most important people in my life, i.e., my family. This book has been my long-cherished dream which would not have been turned into reality without the support and love of these amazing people. They have encouraged me despite my failing to give them the proper time and attention. I am also grateful to my best friends, who have encouraged and blessed this work with their unconditional love and patient.

Jyotir Moy Chatterjee
Department of IT
Lord Buddha Education Foundation
(Asia Pacific University of Technology & Innovation)
Kathmandu, Nepal

Writing a book is harder than I thought and more rewarding than I could have ever imagined. First and foremost, I would like to thank my father Mr. Krishan Dev Pandey for being coolest father ever and my mother Mrs. Veena Pandey for allowing me to follow my ambitions throughout my childhood. They taught me discipline, tough love, manners, respect, and so much more that has helped me succeed in life. Also, my gratitude to my elder sister Mrs. Arpna Tripathi, who always stood by me during every struggle and all my successes. She has been my inspiration and motivation for continuing to improve my knowledge and move my career forward. Also, I'm eternally grateful to my wife Mrs. Kajal Pandey for standing beside me throughout my career and writing this book. I also thank my wonderful son Aarudra Pandey, for always making me smile and for understanding on those weekend mornings when I was writing this book instead of playing games with him. I hope that one day he can read this book and understand why I spent so much time in front of my computer. Last but not the least, I want to thank my friends who always backed me in my good or bad days and everyone who ever said anything positive to me or taught me something. I heard it all, and it meant something.

Abhishek Kumar
Department of Computer Science & Engineering
Chitkara University Institute of Engineering and Technology
Chitkara University, Punjab, India

CHAPTER 1

Introduction to Blockchain Technology

GEETA NIJHAWAN[1*], SHRUTI VASHIST[2], ANITA KHOSLA[1], and
SIDDHARTH SAGAR NIJHAWAN[3]

[1]*MRIIRS, Faridabad, Haryana, India*

[2]*Manav Rachna University, Faridabad, Haryana, India*

[3]*Netaji Subhas University of Technology, New Delhi, India*

Corresponding author. E-mail: geeta.fet@mriu.edu.in

ABSTRACT

The 21st century has seen tremendous growth in terms of technology. In order to modernize and make life comfortable, people have widely accepted these technological advancements. Lately, technologies like controlling devices using remotes and using voice to give commands have become part of modern life. We have seen significant growth in the field of augmented reality and Internet of things (IoT) in the past decade. The recent addition to the list is blockchain technology. A blockchain can be described as a growing list of records, called blocks. These are linked using cryptography. Each block of blockchain contains a time stamp, transaction data and a cryptographic hash of the previous block. It is not possible to modify data in a blockchain.

Blockchain-based applications are fast growing up. It can be used in financial services, education system, healthcare, IoT, and many more. However, there are many challenges of blockchain technology which need to be addressed. The major ones are scalability and protection problems. The chapter gives an overview of the blockchain technology. The basic concepts and working of blockchain are discussed in detail. The challenges and applications of blockchain technology are also presented. Lastly, the future trends for blockchain are discussed.

1.1   INTRODUCTION

At present, the transaction of money between two parties is handled by third party. The control is not in the hands of the parties actually involved in the money transaction. If we want to make payment through digital mode then we need a middleman which can be a bank or credit card company to complete the transaction. The transaction is centralized and many a times we have to pay transaction fee to the bank or service provider. The same scenario exists in case of games, music, software, etc. This issue is addressed by blockchain technology. It creates a decentralized environment in which the third party is not in control of the transactions and data. Blockchain can be defined as a distributed database solution that keeps a record of continuously growing list of data records. The transactions are confirmed by the nodes participating in it. A public ledger is used to record all data which includes information of every completed transaction. In blockchain, the information regarding every completed transaction is shared and available to all nodes. Hence, it can be said that blockchains are more transparent than the third-party systems. Also, all the nodes in blockchain are anonymous. This makes the system all the more secure. Figure 1.1 shows the three types of systems, that is, centralized, decentralized, and distributed.

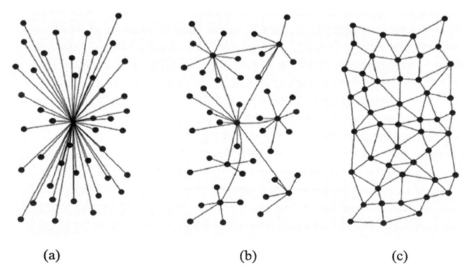

| (a) | (b) | (c) |

FIGURE 1.1   Evolution of networks: (a) centralized, (b) decentralized, and (c) distributed networks.

In centralized systems, several jobs are done on a particular central processing unit (CPU) whereas in decentralized systems, every node makes its own decision. The final behavior of the system is the aggregate of the decisions of the individual nodes. In distributed systems, jobs are distributed among several processors. The processors are interconnected by a computer network.

Architecture of Distributed System

- Peer-to-peer: all nodes are peer of each other and work toward a common goal.
- Client–server: some nodes become server nodes for the role of coordinator, arbiter, etc.
- N-tier architecture: different parts of an application are distributed in different nodes of the systems and these nodes work together to function as an application for the user/client.

The application that launched blockchain technology for the first time was bitcoin. A decentralized network was created by bitcoin for cryptocurrency, where the participants can make transactions with digital money. However, there are still some technical issues with blockchain that required to be addressed. The security and privacy of nodes is required in order to prevent attacks by hackers which may disturb transactions in blockchain (Swan, 2015). Also, some amount of computational power is required while confirming transactions in the blockchain. To address these questions, this chapter presents a systematic introduction to blockchain and its working (Kitchenham and Charters, 2007).

The chapter is organized as follows: The background of blockchain and bitcoin is given in Section 1.2. Also, the technical aspects of blockchain technology are presented by discussing its history, structure, and functionality. Sections 1.3 and 1.4 deal with main characteristics and the working of blockchain. Real-world applications are discussed in Section 1.5. The limitations and challenges of blockchain technology are dealt in detail in Section 1.6. The conclusions are given in the end (Antonopoulos, 2007).

1.2   BASIC CONCEPTS OF BLOCKCHAIN

Until the blockchain was invented, it was not possible to do individual transactions over the Internet (Yli-Huumo et al., 2016). It required centralized monitoring and control to have non repudiation or disapproval of data. It was difficult to build trust between different parties. There were chances that one of the parties can modify the data for their advantage without the knowledge of the other party. A group consisting of distributed individuals required a centralized authority to verify their transactions. This situation is similar to the "Byzantine Generals Problem" (Wright and Filippi, 2015). This provided a solution to the problem that it is possible for distributed computers to make a decision against an attack from unwarranted parties without any centralized control (Boucher et al., 2017). It was assumed that Byzantine army has been divided into three parts and they are planning an attack at the enemy city. The leaders of three divisions are independent to take decisions but in order to take control of the city they have to reach a common consensus. The leaders can send communications through messenger only, but in between a traitor is trying to ruin the leaders' moves to reach a common consensus so that they cannot attack the enemy unitedly. He does so, either by misguiding the leaders to attack the enemy before others or hiding some significant information (Wright and Filippi, 2015). One way to avoid tampering of data is to encrypt it before sending the message. A probabilistic approach is being used by blockchain for providing a solution to the problem of Byzantine Generals (Nakamoto, 2008). The transparency and reliability of data is increased when it moves through a network of computers. Hence, the chances of unauthorized attack are reduced significantly and it becomes difficult to manipulate a distributed database with spurious data. There is only one way to attack and that is when the attacker uses more computational power than that of the entire network. The protocols used in blockchain technology ensure that transactions are done correctly (Christidis and Devetsikiotis, 2008).

Imagine that everyone in the world one day decide that they will trade with a new currency. They do not want banks and their governments to have any control in this currency. The currency should be truly universal in all sense and making it digital would allow for the fastest transfers possible. Such a system was indeed created in 2009 by an anonymous person(s) called Satoshi Nakamoto. Bitcoin is the world's first digital cryptocurrency. It does not have any central authority exercising control. Blockchain is the concept and bitcoin are the implementation (Nakamoto, 2008). The blockchain can be described as a chain of connected blocks where each block contains information about transactions. Each block contains the hash, time stamp and transactional data of the previous block (Fig. 1.2). When a block completes the transactions, a new block gets created. It has the same properties and hash as that of the preceding block (Crosby et al., 2016).

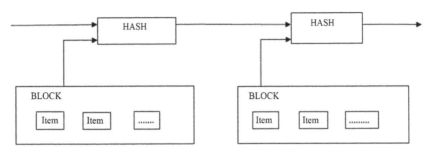

FIGURE 1.2 Connection of blockchain's blocks (Holotescu, 2018).

The blockchain contains a database of transactions, which are recorded into blocks in chronological order and verified by other computers in the network. Also, each block is defined by a complex mathematical equation also called hash functions, which are responsible for data integrity and nonrepudiation (Holotescu, 2018). A copy of blockchain's database is available with each participant in the network. The synchronization of the computers takes place at regular intervals so that all participants get the correct and original copy of the shared database. All transactions which occur in currency are recorded in the copy of blockchain. Blockchain participants are known as nodes. Each node has a unique network address at any timeframe. There is only one route or chain to the first block from any random block (Sultan, 2018). In a case when the transactions are added and a few other blocks are created at almost same time, a node is built on the block that is received first. The block that is added first is included in the main chain, because that is the longest chain at that moment. However, there might be several chains or forks from the initial block (Fig. 1.3).

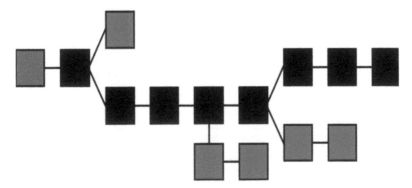

FIGURE 1.3 Main and forked chains into blockchain.

In Figure 1.3, the main chain is shown in black. It has the maximum number of blocks from the origin to the current block (Kehrli, 2016). The first block is represented in green. The purple colored side blocks also called forks. They have no relation to the main chain. The nodes that complete the consensus mechanism process in the bitcoin application are called miners. In consensus mechanism process, the transactions contained in a block are first verified, and then the blocks are published. The blockchain are decentralized systems, and provide safe and secure systems because transactions are encrypted before storage. The process of blockchain is described below (Kehrli, 2016):

- For blockchain transactions, we don't use a central clearing authority. Instead, all computers (nodes) participating in blockchain network share an open ledger that contains ALL transactions that have taken place for every account so far.
- Each new transaction is broadcast to all available nodes for verification.
- If a majority of nodes approve the transaction, it is considered valid and added to the ledger (Khudnev, 2017)

Blockchain contains a number of blocks strung together (Pilkington, 2016). However, during the process of addition of a block following steps are involved:

1. There should be a transaction.
2. Verification of transaction takes place.
3. The transaction is required to be stored in a block.
4. The block is then given a hash.

The working of blockchain is shown in Figure 1.4.

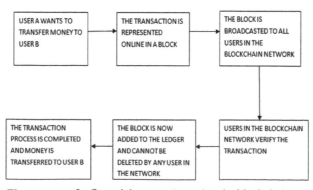

FIGURE 1.4   The process of a financial transaction using the blockchain technology.

A block consists of four details:

Hash of previous block: It holds the hash value of the previous block.

Transaction data: Contains details of several transactions.

Nonce: A nonce is a random value which is used to vary the value of hash.

Hash: A hash is an alphanumeric value which is used to identify a block.

The block gets publicly accessible as soon as it is added to the blockchain. No one can erase the transaction, and because it cannot be erased, no one has to trust an authority to keep it safe. Everyone has the information so "cheating the system" is impossible (Kitchenham and Charters, 2016).

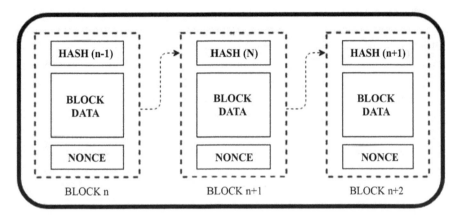

FIGURE 1.5 Connection of blocks.

1.3 WORKING OF BLOCKCHAIN

All over the world, there are a number of people possessing bitcoin. It is a digital currency which is used to send and receive money across the world in a decentralized manner with no transfer fees. Bitcoin does this by storing and transacting the money over a distributed peer to peer network called Blockchain (Li et al., 2017).

When we make transactions with printed money, it is monitored and authenticated by a central authority like a bank. But bitcoin transactions have no third-party interference.

When any person does transactions, that is, he makes payment to other person for his purchase using Bitcoin, the transaction is verified by the computers on the network. To accomplish this, a complex mathematical problem called a "hash" is required to be solved, by running a program on their computers (Zheng, 2017). When the transaction gets completed the block gets publicly stored on the blockchain, and then it cannot be altered. A computer that completes the task successfully gets reward for the labor in terms of cryptocurrency (Peck, 2017).

It is to be noted that even though transactions are publicly recorded on the blockchain, user data is kept secret. In case of bitcoin, the transactions are conducted when participants successfully complete a program called a "wallet." Every wallet has two cryptographic keys called a public key and a private key. These are unique and distinct. Public key is used while depositing and withdrawing the transactions. It also appears as the user's digital signature on the blockchain ledger.

A user needs private key in order to withdraw a payment (Boucher et al., 2017). The public key of user is generated through a complicated mathematical algorithm and is an abridged version of their private key. This is created using complex equations. The reverse process is almost impossible and one cannot generate a private key from a public key. That's why blockchain technology is considered confidential (Peters and Panayi, 2016).

The different types of blockchains are shown in Figure 1.6. There are two types of blockchains: permission less (public) and permissioned (private). Permissions less blockchain networks are used in most of the market's digital currencies. Every user creates a personal address and then interacts with the network. They can submit transactions; and can add entries to the ledger. Permissioned blockchains are not open systems. The users cannot join the network freely. In order to view the recorded history, permission is required. They cannot do transactions on their own (Sikorski et al., 2017).

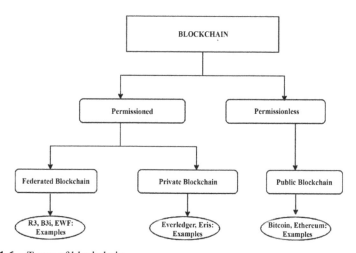

FIGURE 1.6   Types of blockchain.

Public Keys and Private Keys

To understand the difference between these two better let us assume, a public key is similar to a school locker and the locker combination can be called as private key. The locker is accessible to all teachers and students. Anyone can put letters or notes in the locker without seeking permission for it. However, in order to retrieve the contents of the mailbox, a unique key is required. This can be done only by an authorized user (Sultan, 2018). Although these lockers are safely kept at a central place there is no single person who has record of school locker combinations. Similarly, there is no central authority or database that has the record of private keys of a blockchain network. The user will not be able to access their wallet if they misplace their private key. Federated blockchain removes the sole autonomy of any one entity by using private blockchains. So, instead of one in charge, there may be more than one in charge (Stanciu, 2017).

In the bitcoin network, the copy of the blockchain is available with all the computers. It gets updated upon addition of a new block. But sometimes it happens that because of some fault or by the ill intentions of a hacker, a copy of blockchain available with a user may become different from other copies of the blockchain (Autonopoulous, 2014). Hence, it is difficult to tamper the data on blockchain network.

Multiple blockchains cannot exist. The blockchain protocol uses a mechanism called "consensus." In case there are several and different copies, the blockchain protocol will select the longest chain. If there are a greater number of users, it means that we can add blocks quickly toward the end of the chain. Hence, the blockchain will be the one that is trusted by maximum number of users. The consensus protocol is considered to be the greatest strength of blockchain technology but may become its weakness too (Crosby et al., 2018).

Theoretically, a hacker can use 51% rule which is called majority attack. Let's assume there are about 5 million computers on the Bitcoin network. An attacker needs to control minimum of 2.5 million plus one computer so as to get a majority. If the hacker succeeds, it is possible to interfere with the recording process of new transactions. They can reverse a transaction and it appears to the user that still has the coin they had spent. This is called double-spending. It is a fraud involving digital currency and causes users to spend their bitcoins twice.

In short, the following steps are involved during transactions in blockchain technology (Holotescu, 2018):

1. The network uses public and private keys so as to form digital signature for authentication.
2. The next step is authorization.
3. Verification is done through complex mathematical calculations so as to reach a consensus.
4. To make transactions, the private key of sender is used and the transaction information is announced over the network. It creates a block which contains data regarding digital signature, time stamp, and the public key of receiver.
5. The validation process starts after the block of information is broadcasted.
6. In order to process the transactions, mathematical puzzle is required to be solved by miners. Mining is the process of adding transactions. The miners are required to invest their computing power to solve these problems.
7. The miner who solves the puzzle first gets bitcoins as incentives. These problems are known as proof-of-work.
8. Consensus protocol is used to add a block to the existing blockchain.
9. After a new block is added to the chain, the existing copies of blockchain are updated.

1.4 BLOCKCHAIN CHARACTERISTICS

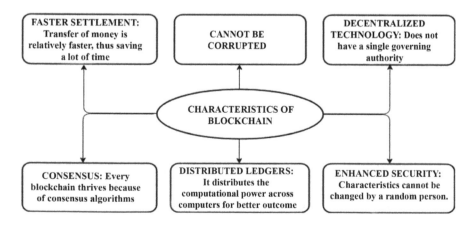

FIGURE 1.7   Characteristics of blockchain.

The features of blockchain technology are discussed below (Singh and Singh, 2016; Tschorsch, 2016):

Decentralized Computing: Blockchains are decentralized in nature which means there is no centralized control. The authority of the overall network is not with a single person or a group. Although the copy of the distributed ledger is with all the participants in the network, but still they cannot modify it on their own. This feature of blockchain makes system more transparent and secure while giving power to the users (Yli-Huumo et al., 2016).

Immutable: The immutability property of a blockchain means that data which has been written once on the blockchain cannot be changed. To make clear the concept of immutability, let us consider that an email has been sent. We cannot take it back once it has been sent. The only possibility is that we ask all the recipients of the email to delete it. This is very difficult task. This is how immutability works (Yue, 2016).

It is impossible to alter or change the data on blockchain after it has been processed. If the data of one block is changed, then it is required to change it in all the succeeding blocks. This is because each block contains the hash of its previous block. As discussed earlier, hash function of a block is unique. If one hash is changed then it requires that all the following hashes to be changed. It is very difficult to change so many hashes because enormous computational power is required to do so. Thus, the data once stored cannot be altered and it is immune to hacker's attack (Zheng et al., 2017).

The immutability property makes it possible to detect tampering of data stored on blockchains. Blockchains can be said to be tamper-proof because changes in any block can be detected and addressed easily (Khudnev, 2017).

Security in Blockchain: In bitcoin blockchain, the identity of the users who are involved in transactions is not disclosed. But in blockchain, the transactions are not done with anonymity. One can have access to personal information involving user's digital signature or username.

This is a cause of concern that if we are not aware of the user who is adding blocks to the blockchain, it is difficult to trust blockchain or the network of computers (Peck, 2017; Li et al., 2017). Blockchain technology addresses the problem of security and trust in a big way. The new blocks are added in a linear and chronological order. They are appended at the "end" of the blockchain and after the addition of a block, it is very difficult to go back and modify the data stored in the block. Each block carries a hash function and also the hash of the previous block. A mathematical function is used to create hash codes. If someone tries to edit the information by any way, it causes a change in hash code as well.

This is very important to security. Suppose a hacker tries to edit transaction resulting in the user to pay for the purchase twice. When the dollar amount of the transaction is edited by the hacker, the hash of the block will change. The hacker would have to modify the hash of the next block which contains the old hash. This causes a change in that block's hash. If a hacker wants to change a block, then it is required to modify every block succeeding it on the blockchain. Modifying so many hashes requires a lot of computing power. Hence, in blockchain technology it is not only difficult but almost impossible to edit or delete a block once it has been added to the network.

The issue of trust is addressed in blockchain networks by carrying out tests for computers that wish to add blocks to the chain. These tests are known as "consensus models." In order to participate in a blockchain network the users are required to "prove" themselves. A common example used by bitcoin is "proof of work."

Computers are made to solve a complex computational math problem so as to "prove" that "work" has been done by them. A block can be added to the blockchain if the computer solves one of these problems. The addition of blocks to the blockchain, called "mining" in the world of cryptocurrency is a tedious process. Actually, the chances that these complex problems will be solved are about 1 in 7 trillion on the bitcoin network. It requires significant amount of power and energy to run those programs on computers. The attacks by hackers are made almost impossible by proof of work. It also makes them somewhat useless.

Efficiency: It takes few days to settle the transactions made through central authority. We must have experienced that a cheque deposited on Friday gets credited to our account not before Monday. The financial institutions like banks etc have fixed business hours, that is, 5 days a week, but blockchain is on work for 24 h a day and 7 days a week. It is possible to complete the transactions in about 10 min. It gets secured within few hours of transaction. This is very beneficial for trades between different countries, which may take long time due to different time zone and also it is required that all parties must confirm payment.

1.5   HOW CAN BLOCKCHAIN BE USED IN THE REAL WORLD

Blockchain technology has proved beneficial not only for monetary transactions but it has proved to be very efficient way to store data about different types of applications also. The blockchain technology can be very reliable for payments, supply chain monitoring, healthcare systems, property exchanges, voting systems, etc. The various applications of blockchain are being discussed below (Boucher et al., 2017).

Banks

Banking sector has been the most benefitted one due to integration of blockchain into business operations. If we make transactions through banks it takes few days to credit the money because of the huge workload on banks (Peters and Panayi, 2016). The transactions can be processed by blockchain in less than 10 min. The time required to complete the transaction is just the time required for adding a block to the blockchain. With the help of blockchain, banks can also exchange funds in a more quick and secure manner between institutions. In case of stock markets, it may take up to 3 days for settlement and clearing process. It sometimes takes even more time if the banks are international. This means that both the money and the shares get blocked for that duration.

Considering the amount of money involved, it carries significant costs and risks for banks for the duration the money is in transit. We can save each year a lot in fees incurred in banking and insurance transactions if blockchain-based applications are used.

Digital Cryptocurrency

Blockchain is the concept for cryptocurrencies like bitcoin. Bitcoin is the most popular example of a blockchain technology (Antonopoulos, 2014). It is a disruptive technology which means that it will change the way the banks are operating with the help of a single centralized secure system. It is going to basically provide an alternative, wherein transactions are kept anonymous and "distributed consensus" are basically used. So, bitcoin has been successfully implemented with the help of a blockchain. It has all security provisions which are otherwise possible only through the centralized banks (Pilkington, 2014).

If a person's bank collapses or there is an unstable government in a country, there are chances that the value of their currency may fall. Cryptocurrencies reduces risk as there is no central control. Many of the processing and transaction fees are not charged in this case. This is more stable form of currency. It provides a larger network of individuals and institutions with whom one can do business with, on both domestic and international fronts.

Healthcare

People in healthcare sector can bank upon blockchain to keep their patients' medical records securely. Blockchain in healthcare, has the potential to tackle the issues of security and data integrity of the large amount of patient data handled by doctors, hospitals, and insurance companies like as immutability, decentralization, and transparency, which can address issues like incomplete records at point of care and difficult access to patients' own health information. It provides confidence to the patients that the record cannot be tampered with. The patient's personal health records are concealed with a private key and then stored on the blockchain (Yue et al., 2016). This ensures privacy and is only accessible only to certain individuals.

Property Records

The process to record property rights is cumbersome and not efficient. At present, the physical deed is to be delivered at the local registration office to a government employee. The entry is done manually into a central database and allotted a registration number. If there is some dispute the claims to the property are resolved with the help of registration number (Holotescu, 2018). This is not only expensive and time-consuming process but it is also inefficient. Sometimes human error may happen which makes tracking property ownership difficult. Blockchain can efficiently handle this issue by eliminating the requirement of searching property documents and tracing physical files in the registration offices. If the record ownership of property is stored on the blockchain, this builds a trust amongst owners that their property deed is safe and permanent (Stanciu, 2017).

Internet of Things (IoT)

The IoT is experiencing exponential growth in research and industry, but it still suffers from privacy and security vulnerabilities. Conventional security and privacy approaches are not inapplicable for IoT, mainly due to its decentralized topology and the resource constraints of the majority of its devices. Blockchain that uses the cryptocurrency bitcoin can be used to provide security and privacy in peer-to-peer networks with similar topologies to IoT (Kehrli, 2016).

Smart Contracts

A smart contract is a computer code which is built into blockchain to facilitate, verify, or negotiate a contract agreement. A set of conditions are made for the operation of smart contracts to which all users agrees (Christidis and

Devetsikiotis, 2016). Whenever these conditions are satisfied, automatically the terms of the agreement are carried out. Consider a case, where an owner wants to rent his apartment using a smart contract. The owner makes an agreement with the tenant that he will share the apartment's door code on receipt of security deposit. Both the owner and tenant would be required to send their share of the deal to the smart contract. The contract will hold and the door code will be shared automatically when the security deposit is received. The smart contract refunds the security deposit in case the door code is not supplied by the rental date. This does not require fees to be paid to notary or third-party mediator (Christidis and Devetsikiotis, 2016).

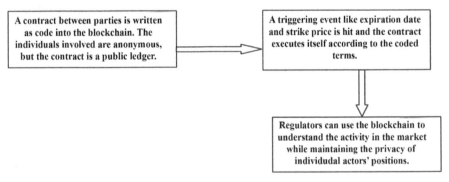

FIGURE 1.8   The process of smart contract.

Supply Chains

The blockchain can assist in the following ways:

1. **Collection of** product and supplier data, from source to store.
2. **Verification of** supplier and product information worldwide.
3. **Improvement of** supply chain compliance, analyze data, and increase security with blockchain.
4. **Communicate** key product and origin information via smart mobile apps.

Voting

Blockchain can be used for voting. It can reduce fraud and booth capturing. It can increase the faith of voters and gives boost to voter turnout. The votes cannot be tampered with once they are stored on blockchain. The blockchain protocol can ensure transparent electoral process. It can reduce the manpower required for conducting election and also the results can be declared instantly.

1.6   CHALLENGES OF BLOCKCHAIN

Although there are numerous advantages of using blockchain technology, but some major challenges need to be overcome before its widespread acceptance (Ahmad and Salah, 2018). The major hurdles are not just technical toward the adoption of blockchain technology. There are lack of standards and regulations. It requires thousands of hours of human manpower and money for design of software and back-end programming which is needed for integration of blockchain to existing business networks. Major challenges which come in the way of acceptance of blockchain technology are discussed below:

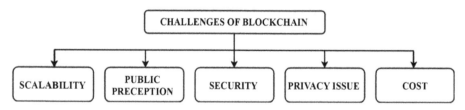

FIGURE 1.9   Challenges of blockchain technology.

Cost

We have discussed that transaction fees are not charged in case of blockchain but the technology does not come free. To validate transactions, bitcoin uses the "proof of work" system. The bitcoin requires appreciable amount of computational power to validate transactions (Wright and De Filippi, 2015). This energy is far from free and costs money. In addition to the cost incurred in mining bitcoin, the user has to pay for electricity bills. Large amount of electricity is consumed while validating transactions on the blockchain. Miners are rewarded with bitcoins to compensate for the time and energy when they add a block to the bitcoin blockchain (Asikorski et al., 2017). In case of blockchains where cryptocurrency is not used, however, miners are to be paid or otherwise given some incentive for validation of transactions.

Blockchains will need to use other methods to achieve consensus, such as the proof-of-stake algorithm which requires much less energy.

Scalability

There is need to increase capacity of blockchain. Blockchains are having trouble effectively supporting a large number of users on the network. Both bitcoin and ethereum, the leading blockchain networks, have experienced slowed transaction speeds and higher fees charged per transaction as a result of a substantial increase in users. The "proof of work" system in bitcoin requires about ten minutes in order completing addition of a new block to the blockchain. Hence, the blockchain network can process only seven transactions per second (TPS). There are other cryptocurrencies like ethereum and bitcoin cash which process 20 TPS and 60 TPS, respectively, but their use is still limited by blockchain. Legacy brand Visa has the capability to do 24,000 TPS.

Privacy in Blockchain

The bitcoin blockchain is designed to be publicly visible. All the information pertaining to a transaction is available for anyone to view. While this feature may be important in some contexts, it becomes a liability if distributed ledgers are to be used in sensitive environments. For instance, private patient data should not be available for all as is the case with proprietary business data. This is also applicable to government data or financial data (Antonopoulos, 2014).

Security

The blockchain maintains confidentiality to protect users from hackers and hence provides privacy. But the blockchain network can be used for illegal activities and trade purposes. Users who make online transactions using blockchain are not given complete anonymity as per current US regulation. In the United States, when a user opens an account, they are required to share information, the identity of each customer is verified and it is checked that the customer does not belong to any known or suspected terrorist organizations.

Investigations have been carried out by several central banks into digital currencies. It is required to make a fool-proof system so that distributed ledger technology could be used without compromising with central bank's ability to control its currency. It is also required to protect the system against any systemic attack."

Blockchain networks and newer cryptocurrencies are prone to 51% attacks. It requires enormous amount of computational power to have control over majority networks of a blockchain and hence these attacks are almost impossible to execute but latest research has shown that this might change. About 51% attacks are likely to increase as it is possible for hackers to simply rent computational power to get majority control (Stanciu, 2017).

For this reason, bitcoin mining pools are monitored closely by the community, ensuring no one unknowingly gains such network influence.

Public Perception

Perception, blockchain holds in the eyes of people is the biggest drawback in the way of its success. Firstly, people have not accepted it as a part of mainstream functioning. Secondly, most of the people are sceptical about the longevity of this technology. The lack of governance and regulation, easy access to become a member of public blockchain adds to the deterioration of image of blockchain in the eyes of people. All these factors are the hurdles for the growth of this technology.

1.7   CONCLUSION

Blockchain is still at its nascent stage of research and development. Researchers working in the domain of security and cryptography are working hard to take it further to newer heights. This technology can be very beneficial for both financial and nonfinancial sectors. The issues of reliability, security, and shared knowledge can be taken care of at the same time (Khudnev, 2017).

Since its inception, it has been one of the most attractive technologies. Hence, it can be said that that there are numerous opportunities of research in this area and it is the need of the hour to explore and improve the blockchain technology. This can be done by minimizing the flaws and by enhancing its efficiency.

KEYWORDS

- **blockchain**
- **cryptography**
- **Internet of things (IoT)**
- **end-to-end encryption**
- **transaction**

REFERENCES

Antonopoulos, A. M. Mastering Bitcoin: Unlocking Digital Cryptocurrencies. O'Reilly Media, Inc., 2014.

Boucher, P.; Nascimento, S.; Kritikos, M. How Blockchain Technology Could Change Our Lives', European Parliamentary Research Service. www. europarl. europa. eu/RegData/ etudes/IDAN/2017/581948/EPRS_IDA. 2017.

Christidis, K.; Devetsikiotis, M. Blockchains and Smart Contracts for the Internet of Things. *IEEE Access* **2016**, *4*, 2292–2303.

Crosby, M.; Pattanayak, P.; Verma, S.; Kalyanaraman, V. Blockchain Technology: Beyond Bitcoin. *Appl. Innov.* **2016**, *2* (6–10), 71.

Holotescu, C. Understanding Blockchain Technology and How to Get Involved. In *The 14th International Scientific Conference eLearning and Software for Education Bucharest*, 2018; pp 1–8.

Kehrli, J. Blockchain Explained. Netguardians (en línia). (Data de consulta: 25 de juny de 2017) 2016. https://www.netguardians.ch/news/2016/11/17/blockchain-explained-part-1.

Khan, M.; Salah, K. IoT Security: Review, Blockchain Solutions, and Open Challenges. *Future Gener. Comput. Syst.* **2018**, *82*, 395–411. DOI: 10.1016/j.future.2017.11.022.

Khudnev, E. Blockchain: Foundational Technology to Change the World. *Int. J. Intell. Syst. Appl.* **2017**.

Kitchenham, B.; Charters, S. Guidelines for Performing Systematic Literature Reviews in Software Engineering; 2007.

Li, X.; Jiang, P.; Chen, T.; Luo, X.; Wen, Q. A Survey on the Security of Blockchain Systems. *Future Gener. Comput. Syst.* **2017**.

Nakamoto, S. Bitcoin: A Peer-To-Peer Electronic Cash System. http://bitcoin.org/bitcoin Pdf, 2008.

Peck, M. E. Blockchain World—Do You Need a Blockchain? This Chart Will Tell You if the Technology Can Solve Your Problem. *IEEE Spectr.* **2017**, *54* (10), 38–60.

Peters, G. W.; Panayi, E. Understanding Modern Banking Ledgers Through Blockchain Technologies: Future of Transaction Processing and Smart Contracts on the Internet of Money. In *Banking Beyond Banks and Money;* Springer: Cham, 2016; pp 239–278).

Pilkington, M. 11 Blockchain Technology: Principles and Applications. In *Research Handbook on Digital Transformations*, 2016; p 225.

Sikorski, J. J.; Haughton, J.; Kraft, M. Blockchain Technology in the Chemical Industry: Machine-to-Machine Electricity Market. *Appl. Energy* **2017**, *195*, 234–246.

Singh, S.; Singh, N. Blockchain: Future of Financial and Cyber Security. In *2016 2nd International Conference on Contemporary Computing and Informatics (IC3I)*, 2016; pp 463–467. IEEE.

Stanciu, A. Blockchain Based Distributed Control System for Edge Computing. In *2017 21st International Conference on Control Systems and Computer Science (CSCS)*, 2017; pp 667–671. IEEE.

Sultan, K. Conceptualizing Blockchain: Characteristics & Applications. In *11th IADIS International Conference Information Systems;* Ruhi, U., Lakhani, R., Eds.; 2018; pp 49–57.

Swan, M. Blockchain: Blueprint for a New Economy. O'Reilly Media, Inc. , 2015.

Tschorsch, F.; Scheuermann, B. Bitcoin and Beyond: A Technical Survey on Decentralized Digital Currencies. *IEEE Commun. Surv. Tutor.* **2016**, *18* (3), 2084–2123.

Wright, A.; De Filippi, P. Decentralized Blockchain Technology and the Rise of Lex Cryptographia. SSRN 2580664; 2015.

Yli-Huumo, J.; Ko, D.; Choi, S.; Park, S.; Smolander, K. Where is Current Research on Blockchain Technology? A Systematic Review. *PloS One* **2016,** *11* (10), e0163477.

Yue, X.; Wang, H.; Jin, D.; Li, M.; Jiang, W. Healthcare Data Gateways: Found Healthcare Intelligence on Blockchain with Novel Privacy Risk Control. *J. Med. Syst.* **2016,** *40* (10), 218.

Zheng, Z.; Xie, S.; Dai, H.; Chen, X.; Wang, H. An Overview of Blockchain Technology: Architecture, Consensus, and Future Trends. In *2017 IEEE International Congress on Big Data (Big Data Congress)*, 2017; pp 557–564. IEEE.

CHAPTER 2

An Overview of Blockchain Technology: Architecture, Consensus Algorithm, and Its Challenges

POOJA DIXIT[1], ANUJA BANSAL[2], PRAMOD SINGH RATHORE[3*], and MANJU PAYAL[4]

[1]Dezyne Ecole College, Ajmer, India

[2]Dr. Ambedkar College, Tonk, Rajasthan, India

[3]ACERC, Ajmer, Rajasthan, India

[4]Academic Hub, Ajmer, India

*Corresponding author. E-mail: pramodrathore88@gmail.com

ABSTRACT

Blockchain technology is a public ledger which is factually needed in distributed system that records all transactions or digital events that have be executed sharable among participating parties in distributed network. Each transaction is verified by consensus algorithm techniques and allow for transfer records in participating network. Once information is entered it can never be altered or erased. Thus, blockchain contains definitive and verified records of every single record. Consensus algorithm plays a vital role in blockchain technology, which decides the agreement about each transaction and append new block in transaction according to agreement. Bitcoin is a first emerging technology which firstly blockchain technology. Bitcoin is a decentralized, peer-to-peer digital cryptocurrency. However, this digital currency is controversial technology in which blockchain technology work flawlessly and widely used in both financial and nonfinancial applications. The reason behind uses of blockchain technology is its security, anonymity, its central attributes, and data integrity without any interference of third party. This chapter presents an extensive overview of blockchain technology with its consensus algorithm. This chapter also compares some typical consensus algorithm used in blockchain. Furthermore, some technical challenges and functional requirements, and some possible future trends are briefly listed.

2.1 BLOCKCHAIN TECHNOLOGY

History of blockchain: Today, blockchain is a revolutionary technology. It is distributed ledger that stores all transaction records between two parties efficiently and securely. The history of blockchain starting with the discussion with white paper bitcoin which was released by Satoshi Nakamoto's in 2008 titled "Peer to Peer network." The hypothesized of this paper was to direct online payment in between two parties securely without any one of third party or intermediary. In simple, the paper describes the online digital transaction that uses cryptographic technique instead of trust. Blockchain technology runs in between Bitcoin. Bitcoin is a digital currency introduced in 2009 and in last decade it is increasingly successful. In many sectors, bitcoin is used for transaction. Bitcoin use the basic principle of blockchain. Blockchain is a type of database that contains many duplicate copies of records on multiple nodes or computers. And all these nodes contain same information. So blockchain is a good choice for digital transaction because it keeps all records in public space and doesn't allow anyone to remove them. Each block can contain multiple transactions at a time. Each transaction contains unique reference number, a time stamp, pointer that point previous transaction, and information of transaction. Ethereum proposed blockchain in 2013 "a decentralized platform that runs smart contracts." This paper provide a platform for create cryptocurrency for developing, decentralized applications. It defines that blockchain developers create markets, store registries of debts or promises, move funds according to the instructions like a contract, or will without any interference of third party. Ethereum is a ledger technology that is used in companies for developing new programs with this feature of this technology then attract the attention of corporations like Microsoft, BBVA, and UBS which uses this potentially for smart contract to save time and money or give security. The market value of this ethereum contract platform is now in billions. At current time blockchain technology use the proof of work concept which use more computing power for solve critical problems or for mining. This mining process is used for creating new block, and then these blocks are bundled together when new transaction in initiated (LegalTech News, 2018). Ethereum developers use another approach proof-of-stake which goal similar as proof-of-work (POW) but using different process.

Since currently every computer in a blockchain processes every transaction so it is very slow process. So blockchain scaling is a new innovation that is on horizon now days. This scaled technique increases the speed of process without compromising their security. This scaled process simply finds out the system that is necessary to confirm the transaction and it divide work in equality. So, this makes this process very faster. These all innovations prove the hard work of developers, scientists, and cryptographers that help in blockchain technology. This technique makes everything possible and easier like money transfer, self driving cars, drones, etc. And bitcoin is one of several hundred applications that use this blockchain technique. It has been growing very rapidly in the last decade and considering its facilities, it will grow faster in the coming decade (Bernard, 2017).

2.2   BLOCKCHAIN INTRODUCTION

Today, the cryptography is a major discussion topic in industry or education. From these, bitcoin is a most successful cryptographic technique that achieves almost \$10 billion in capital market in 2016. Which is specially designed for data storage, securely transaction in the betoken network without any interference of third party. Blockchain technology is used in bitcoin cryptography. Blockchain technology is public ledger in which all transactions are stored in a list of blocks, which are continuous chain of multiple block and new block can be added easily. This technology allows transaction securely that use asymmetric cryptography and distributed consensus algorithm. The key characteristics of blockchain technique are decentralization, persistency, anonymity, and auditability. These features make the blockchain very cost efficient time saving or also provide security. some financial services where blockchain technology is mostly used such as digital wealth, online transaction, remittance, and in many other area includes smart contracts, Internet of things, security services, public services, etc. Nature of blockchain technology is unfaltering in which once transaction is initiated cannot be tampered. This feature is used in those business area where require high data reliability and honesty. Besides, blockchain technology is distributed in nature in which if any failure occurs then it doesn't affect the technology.

Blockchain technology is just like a cartulary record. It can be described as chronological transaction of records, which are in the form of blocks. It is growing list of records that are called blocks. Each block is connected with each other in the network by using cryptographic technology that uses mathematical hash function for mapping any size of data. That provides security, data integrity, nonrepudiation of data, and also maintains shared resources. A copy of the blockchain database is stored in each member present in the network. Each computer uses timestamps and synchronizes frequently, ensuring that all connected computers are sharing the original and correct data of the shared database (Zheng et al., 2017).

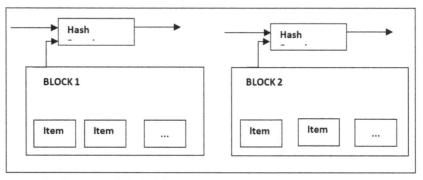

FIGURE 2.1   Structure of blockchain.

Designing purpose of blockchain is instinctively obstructive alteration of data from records. It is an open, distributed ledger of transaction that used in bitcoin. Firstly the principles of blockchain architecture were designed for bitcoin. This was mainly focus for secure and distributed transaction. It is public distributed database that holds encrypted ledgers that can keep record of all transactions between two parties adroitly and in permanent way. Current part of blockchain that records recent transactions is called block; when it is completed then it is permanently stored in blockchain database and a new block is generated.

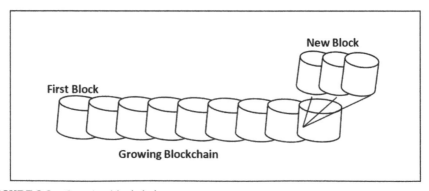

FIGURE 2.2   Growing blockchain.

2.3   OPERATION OF THE BLOCKCHAIN

2.3.1   A SIMPLIFIED VIEW OF THE BLOCKCHAIN

While discussing about the technology of blockchain, one thing that is mentioned is the blockchain, the network which are built around it, and all protocols are involved. These all can be represented under this simplified form:

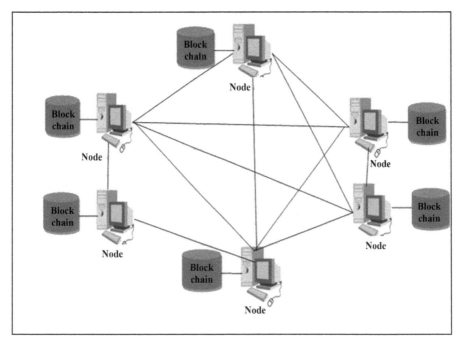

FIGURE 2.3   Simple blockchain network.

The following are the key aspects:

- The network of blockchain is a peer-to-peer network of individual nodes which communicate together through the broadcasting of message.
- Network key component is blockchain and every node of blockchain has its own copy.
- A node should be connected to at least some nodes of them, it is not necessary to be connected with every node.

The blockchain itself is a blocks list. These data "blocks" are digitally recorded and in linear chain it stored. In the chain, every block contains data like bitcoin transaction and it is hashed by cryptography. In the blockchain, the hash is included in every block of the prior block which is linked with two blocks that ensure all data has not been tempered with in the overall "blockchain" and stays unchanged. These have the effect for creating the blockchain from the block of genesis to the block of current. Every block is chronologically ensured to come after the block which is previous because the hash of previous block would not be known. Once a block has been in the chain then it is impractical by computationally for modification because after it each block also has to be restored. The block which is linked with each other takes shape of a chain (Kehrli, 2016).

2.4   THE BITCOIN BLOCKCHAIN

A chain of block is a database of transaction which is shared by every node that is based on the protocol of bitcoin and participating in the system. Every transaction contains the entire copy of blockchain of currency that executed in currency ever. With the help of this information, someone can discover in history that how much value belonged at any point with each address. For every block in the chain has only one path to the block of genesis. However, there can be forks for that which is coming from the block of genesis.

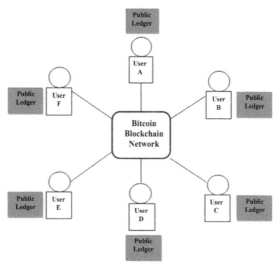

FIGURE 2.4   Bitcoin blockchain network.

One block of forks is created time to time while two blocks are created apart in just a few seconds. The one block which they received first, on that generating nodes are built. Because the chain of blocks is longer so when one block ends up than it is included in the next block and becomes a part of the main chain. By using a protocol of flood on networking, the chain of block is broadcasted for whole nodes.

2.5 OPERATION PRINCIPAL OVERVIEW

For bitcoin transaction, the initial chain of block runs the system of bitcoin, provides distributed secured and public storage. There are some principles of operation to understand:

- If user wants to give some bitcoin to another user then he transmits to the network a transaction.
- As the miners receive the transaction they add it to the current block on which they are presently working on.
- At random one of the miners may "mine" block and win lottery.
- The new "determinate" block is transmitted to network at that moment and then added it to the blockchain everyone's copy (Jerome, 2016).

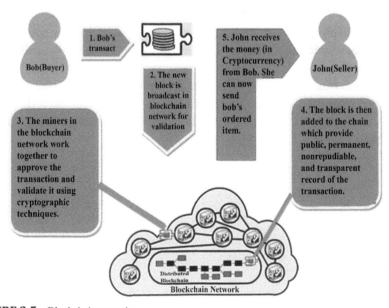

FIGURE 2.5 Blockchain operation.

2.6   TECHNICAL ASPECTS OF THE BLOCKCHAIN

2.6.1   BLOCK ARCHITECTURE

Single block structure: A block is a data structure container. Blockchain brings together an average more than 500 transactions its common size seems to be 1 MB. In bitcoin technology, the size of block can be up to 8 MB. This allows to process one or more transactions in a second. A single block is comprised with header and a long list of transactions (Zheng, 2016).

Block header: Block header stores metadata, which store the information about data which are stored in header. So, the metadata header main purpose is used to identify a particular block in an entire blockchain. It also uses hashed technique to build an authentic work for mining award. The header part of each block comprises with following fields:

- **Index:** It is the first position of the block in blockchain technology. Index of the first block start with '0' and the next '1' and so on.
- **Hash:** Hash function is a mathematical function that enables to fastly identify any data from the database.
- **Previous hash**: Each block in blockchain is connected with its previous block that inherits from its ancestor block. This previous block hash is used to create the new block hash. Each N block contains N-1 hash block.
- **numTx: This specifies the roof transaction in the block.**
- **Timestamp:** It stores the time details when each block created.
- **Nonce:** Stores the random integer (32 or 64 bits) that is used in the mining process.
- **Transaction:** This field stores as an array in the body block which store complete summary of transaction that are performed yet. It uses another data structure for data storage, that's called Merkle trees (Nguyen and Kim, 2018).

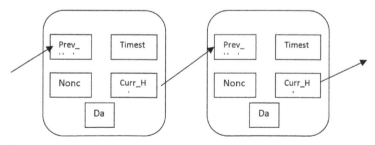

FIGURE 2.6   Example of two subsequent blockchains with their attributes.

Merkle trees: It is a fundamental component of blockchain technology which contains a summary of all transactions in the block. It is a data structure that is used for summarizing data efficiently and also verifies the data in large data sets. Merkle tree is also known as Binary hash tree which contains hash value used in cryptography. Every transaction associated with hash value.

How do Merkle trees work?

Merkle tree is formed with recursively hash pair of nodes until there is only one hash that's called merkle root node. They are constructed from the bottom up, from hashes of individual transactions (known as transaction IDs). It is a binary tree in which all leaf nodes require an even number. In case of odd number of leaf node, last hash will be duplicate one more time to create an even number of leaf nodes.

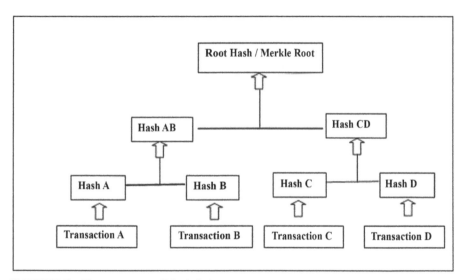

FIGURE 2.7   Tree structure of blockchain.

Source: https://hackernoon.com/merkle-trees-181cb4bc30b4. Reprinted from Ray, 2017.

For example, four transactions in the figure, A, B, C, D use each hash value. The hash is stored in each leaf node that is Hash A,B,C,D. Successive pair of leaf nodes are then summarized in a parent node that refers as Hash A,B and Hash C,D. After that the two hash (Hash A,B and Hash C,D) are hashed again and generate hash that is the merkle root. This process is used with larger datasets and performs hashing until there is only one node at the top. SHA-256 cryptographic hashing algorithm is used to perform this process. It is a secure hash algorithm (Vujicic, 2018).

Block Identifiers

There are two main methods to identify a block, the first is **hashing** and the second is **digital signature.**

- **Hashing** is a function used for convert any input data into an encrypted output of a fixed length. The input can be any number of bits that could be a single character, sound, video files, a spreadsheet, or any type of file which size can be infinitely large. The hashing algorithm can be chosen depending on the needs. Some example of hash algorithms are MD5, SHA-1. An example of bitcoin secure hashing algorithm also called SHA-1. It always returns the fixed length output which is very important to blockchain management in crypto-currency. The main basic part of cryptocurrency is its blockchain, which comprise with global ledger that are linked together with individual block of transaction data. Blockchain allow only validated transactions, which prevents with the fraudulent transaction and also prevent the double spending of currency. This validating process depends on the data encryption using hashing algorithm that's called hash value which use mathematical function to encrypt the data, the resulting encrypted output is a series of numbers and letters that are different of original data. Cryptocurrency mining involves working with this hash.
  Securing Data with Hashing: Using hashing increases the security of the data which cannot be easily accessible by anyone. Anyone who wants access data must decrypt the data. Some properties of cryptographic hash function that makes blockchain data structure functionally sledgehammer.
- Easy to Generate: It is easy to generate hash value by using hash function for any input data and quick to produce hash for any given input data.
- Inconvertible: It is impossible to determine input data based on hash value. This is the most popular quality of hashing for securing data.
- Commitment: Hashing provides data integrity. It is not possible to change the original input text.
- Collision free: It is impossible to use same hash value for two input texts (Blockchain Technology Primer, 2018).

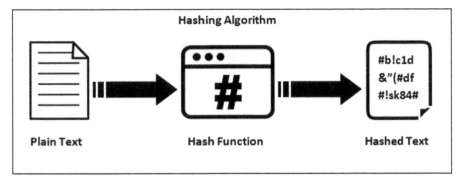

FIGURE 2.8   Hashing.

Source: https://steemkr.com/cryptocurrency/@keysa/for-the-sake-of-clarity-a-lot-of-crypto-terms-that-you-need-to-know

- **Digital signature**: It is an important phase for ensuring the security and integrity of the data that is used on the blockchain. It is a most important standard used in blockchain protocol for secure block transactions, transfer of sensitive information that must not sharable by anyone, software distribution, contract management, and any other situation to detect and prevent any external mess it is important. Digital signature provides three major advantages for the storage and transfer of information on a blockchain. Firstly, it guarantees data integrity. Therefore, encrypted data which are being sent, that can be changed without seeing or any molestation by hacker. If this happens, the signature will also change, which becomes unacceptable. Therefore, digital signatures not only provide data security, but also demonstrate whether it has been tempered with.

For cryptographic transformation of information, it uses asymmetric cryptography, also known as public key cryptography (PKI). It uses public and private keys for encrypted data. The key is simply made up of larger numbers that are paired together and not identical. The key which is shared with everyone is called public key and the other key in pair is kept secret key called private key. It is used to digitally sign messages which are sent to another user. This signature is attached with the message so that the receiver can verify by using sender's public key. In blockchain, every transaction that is executed use digitally signed signature by sender with private key. This signature ensures the integrity of data (Agrawal, 2018).

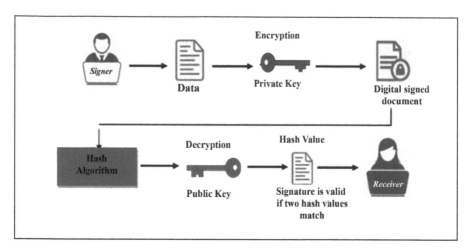

FIGURE 2.9   Cryptographic transformation.

Source: https://www.rozgardesh.com/2018/08/digital-signature.html

Functional characteristics: Blockchain technology can be used independently in different diversity area and markets, and from insurance to the health industry. It can be used virtually, that are used easily in any industry or organization where it managed assets and transaction. It endues a secure chain for both digital and physical assets by using its functional characteristics. These functional characteristics give secure and trustable transaction, consensus, and smart contacts. These features are explored in the following sections (LegalTech News, 2018).

o   **Transactions and smart contracts**: Transaction is a transfer of assets that is broadcast to the entire network and collected in a block that are managed by some entity rules, such rules are operational by using scripting languages like Bitcoin's Forth. This rule is used in advance transactions like multiparty signature, escrow. This rule is configuring with using the smart contracts.

A smart contract is a set of rules or computer code which are build with scripting languages that are embedded into a blockchain to administer the transaction. The main motto of this smart contract is to automatically execute the terms of agreement when the specified conditions are met. In simple words, they are stored programs that are run by the people who developed them.

The benefits of smart contract are more straightforward in business industries collaboration. They provide facilities to verify, apply the negotiation or performance of an agreement or transaction in which they use some type of agreement to use them without any intermediary involvement. It is the simplest form of decentralized automation. This contract is run autonomously to govern the transaction (Buterin, 2016). This contract is just like a smart agent (Stark, 2016). Once it is embedded in the blockchain, it acts as an autonomous agent that can be tamperproof permanently. Then applicant can read, execute, and process the transaction (Sultan, 2018).

o   Consensus and trust: The famous Russian proverb is "Trust, but Verify" is made popular by Regan. The same applies to blockchain. This is contemporary in the case of blockchain where all parties should have similar copies of the block series, but each participant is responsible for verifying with it. Decentralization of blockchain is its core strength, in which a copy of the entire database is accepted by the entire participant. A consensus algorithm is used to ensure the integrity of each copy file. This consensus algorithm ensures that all each data block is legal or not. According to the Coindesk (2017), it also inhibit to attacker that compromising and forking the chain. Nakamoto (2018) suggested a POW approach, which uses a hard cryptographic puzzle that must be solved by miners. POW is the most popular consensus evaluation method. The main aspect of POW is that it is not mandatory for each node to prepare their specific conclusion for a consensus to be reached. It means only one participant in the network can share their conclusion and other node in the network only verify that conclusion (Lisk Academy). Some other consensus model like proof-of-stake, proof-of-burn, proof-of-elapsed-time, and proof-of-capacity are proposed to overcome the weakness of POW model (Kiayias et al., 2017; Zamfir, 2015).

o   The nature of blockchain is reliable and the basic aim is to design for secure and reliable transaction without any third-party involvement. The purpose of this model is to build a trust model which based on the consensus of a group. This model use network for establish and verify any transaction and also provide secure transaction because of no mediator. The thought of trust for this technique is inherent because each block is verified by the group of community which keeps multiple copies of block or record of a blockchain (LegalTech News, 2018).

Perspective of blockchain is trustless. Perception of creating blockchain is to provide secure transaction without any interruption. Traditional database use client–server architecture. Where a client or a user can modify the data which are stored on the centralized server. These data are controlled by referred authority. This authority is responsible for regime of database which gives user authorization to access data from database. If in any case database security is compromised then data from database can be altered or deleted. If in case central authority security is compromised knowingly or inadvertently then blockchain model eliminate the central authority instead of expand the copies of the records to all connected nodes. Each participant's nodes maintain their own copy of blockchain records, and circulate the update copies of records by forming new blocks based on the validated and rules of the consensus model. When each node is validated then it circulates to every node in a blockchain. This process is safer than the traditional centralized model (Sultan, 2018).

Blockchain uses cryptography technology which is based on mathematics. It provides an independent trust for the transaction, and also depends on the consensus model. The recent transaction pool has been kept in the block. Similarly, the blocks are connected with each other by using cryptography block and also verified with a consensus model which comprise with a valuable computing resources (mining).

2.7 PUBLIC AND PRIVATE BLOCKCHAINS

Blockchains can be classified as public, private, or hybrid variants, depending on their application.

o   **Public blockchain:** Public blockchain are permissionless means they are public in nature. Anyone can read, write, or join a public blockchain. In simple words they are decentralized network in which no one control the network and anyone can change and access data anywhere. Bitcoin, ethereum, and litecoin are well known examples that are based on public blockchain (LegalTech News, 2018).
o   **Private blockchain:** A private blockchain is a permission blockchain means networks use restrictions on users who are allowed to participate in the network and in what transactions. In other words, they allow only authorized access of blockchain resources. Participants are known to each other and proven to be trusted. It is a closed

network or centralized network that limits access to certain users. A well-known example would be Hyperledger.

o   **Hybrid blockchain:** This is a combination of both public and private blockchain that assemble both characteristics of both approaches. This is also known as consortium. This approach is only for a consensus privileged group which is controlled by specific privilege servers that are based on consensus rules. Therefore, copies of blockchain are distributed only between authorized parties.

Although public blockchain is based on distributed network that is in the form of decentralized peer-to-peer network. But private blockchain are used in those enterprises where all transaction is recorded and access only within authorized users. Whereas hybrid blockchains are used in small-scale public blockchains. They are in decentralized form, which is accessed by limited participants (Zilavy, 2018).

In summary, blockchain has the following key characteristics.

- **Decentralized systems:** This technology allows user to store data in a system that can be accessible by using network. This data can be anything like document, any contract, etc. Blockchain technology is being viewed as very efficacious decentralized technology. This technique proves their strong feature that massively use in large-scale organizations like banking and finance.
- **Persistency:** This feature allows data persistency that means in case of any invalid transaction then blockchain would not admit it and valid transaction would accept speedily. It is practically impossible to delete and rollback of any transaction. And in case of any block contain invalid transaction then it could be identified immediately.
- **Immutability:** This creates trust in the transaction record. Any transaction in a blockchain is permanent. No one can audit ledger records during transaction. Therefore, it creates trust for securely transaction of records.
- Transparent: Properties of a blockchain are direct therefore anyone can access it or update it. This also provides higher level privacy in which transactions are shared with only those participants who are involved in transaction. This also maintains all tracks about all transactions that can be checked in future (Naidu, 2018).

2.8 BLOCKCHAIN TECHNOLOGY: HOW DOES IT WORK?

By explaining how bitcoin works, we can explain concept of blockchain since it is linked with bitcoin intrinsically. Although, the technology of blockchain is applicable for any insist transaction of digital which exchanged online. Financial institution is tied with the exclusively Internet commerce which serve as the trusted third party that mediates and processes any type of electronic transaction. The trusted third-party roles are to formalize, preserve, and safeguard the transactions. In online transactions, the fraud certain percentage is ineluctable and by financial transaction that need intermediation. The result of this is high cost of transaction (Khatwani, 2018).

For two volitional parties who execute the online transaction over the Internet, bitcoin uses for them the technique of cryptography rather than the trust on third party. Every transaction is protected with a digital signature. Every transaction sent a digital signature using a "public key" of the receiver and sender "private key." The cryptocurrency owner is required to prove their ownership to spend money of "Private Key." The entity which receives the digital currency verified the signature that digital, that is, the "private key" corresponding with the ownership which is used in transaction by using the sender's "public key." In the network of bitcoin every transaction for each node is broadcast and then after the verification it is recorded in ledger of public. Every time before the recording in the ledger of public each one transaction required to be verified to validity. Before recording any transaction the node which is verified needs to ensure two things:

1. The own cryptocurrency of spender, that is, the verification of digital signature on the transaction.
2. Spender has in her/his account sufficient cryptocurrency: in the ledger check each transaction against the account of spender ("public key") to ensure that them has sufficient currency in account.

Although, here is a question of these transactions to maintain the order of it which are transmitted to each node in the peer-to-peer network of bitcoin. There is a requirement of system which ensure that the redundancy of spending cryptocurrency doesn't occur because the transaction is not come in that way in which they generated. It is not ensure that the order of transaction which received on the node are same as they were generated, so we consider that all the transactions transmit node by node through the network of bitcoin. In other words, we can say that there is a requirement to produce a mechanism by which the whole network can concur regarding the transaction order, that is, an intimidating task in the distributed system.

2.9   CONSENSUS ALGORITHMS

What are the consensus algorithms?

Consensus algorithms are a decision-making process used in group. It is a fault-tolerant technique that is used in blockchain systems to obtain the essential agreement in which individual of group make and support the decision. Suppose there is a group of 100 people that want to construct a decision about a project that benefits them all, then any one can suggest an opinion, and then they go with majority that select with consensus. So, it is a method to decide within group. This consensus algorithm not only agrees with majority opinion or vote, but it also benefits all people of group that agree or not with agreement (Zilavy, 2018). This consensus system that is used in agreement is called a consensus theorem. The main objectives of consensus model are:

- Come with an agreement: This technique collects all the agreements from the group more and more as it can.
- Collaboration: Its aim is to make and support the best agreement that help to give best results.
- Co-operation: Each people will work and support the team to accomplish any task, and keep away their own interests.
- Equal Rights: Consensus model is a group task so each people has its own rights to give their votes that means each participant has the same value.
- Actively participation: Participation of individual in a group is necessary. Without any vote and opinion no one can leave the group and also take any decision.

2.10   APPROACHES TO CONSENSUS

Proof-based consensus algorithm: This algorithm is based on POW, POS, and hybrid form. The basic concept of proof-based algorithm is:

Proof of work: It is the first algorithm used in blockchain technology which was proposed by Satoshi Nakamoto in 2009 in the blockchain network that relies on the proof. This blockchain consensus model is used by many block-chain technologies to ensure all the transactions and also making compatible block into the network chain. This algorithm was first used in bitcoin and other cryptocurrencies and later adopted by ethereum. At present this is the only algorithm to prove that it has successfully run against Sybil attacks. The decentralized ledger system gathers all information related to the blocks. This is the responsibility of all individual nodes that's called miners maintaining their information is called mining. The main purpose of this algorithm is to solve all complex mathematical problems and find out the results (Puthal, 2018).

Working: Multiple transactions are rolled up together that's called block. To verify all transactions, mining is used to check the block is legitimate or not. For doing this, miner uses mathematical function or puzzle by using hashing. This is also known as POW which needs more computational power to solve (Puthal, 2018).

Proof of stake (PoS)-based consensus: This model was first suggested on the bitcoin talk forum in 2011. But first, it was used in digital currency by Peercoin in 2012 together with ShadowCash, Nxt, BlackCoin, NuShares/NuBits, Qora, and Nav Coin.

In contrast to POW, which require more computational power to solve mathematical problems, and whose main goal is to validate transaction and create new blocks with the PoS, it is a type of algorithm in which crypto-currency's blockchain network goal is to obtain distributed consensus. In PoS, the creator selects new block by using random combination system and wealth of age. It also focuses on to find that a user holds how many cryptocurrencies, and holds that particular currency how much time. The main advantages of proof of stack are that it uses less energy and gives cost efficient results, give security to validate each block.

Working: This process is based on random selection in which all miners are selected in network randomly. If users have specific amount of coins in wallet then they can be selected randomly and if not have coins then they cannot participate in the staking. So, for being a miner user need to accrue certain amount of coins and after that a voting system is used to choose valuators. This process is fully straightforward where new blocks are created and which is based on proportion to the number of coins in the wallet. For example if the wallet has 10% coins then miners create 10% new blocks.

Hybrid form of PoW and PoS: Hybrid protocol builds together PoW component with a PoS model in bitcoin cryptocurrency. This proof of activity protocol provides good security against unwanted practical attacks on bitcoin. Although in case of possibility of attack use proof of activity because it gives the benefits of both of utilized consensus algorithm. This combination is created for obtain the higher degree of network decentralization and also increase the network level security or its speed. Mining process in Proof of Authority (PoA) starts with PoW where miners try to find new block and outrun each other. They use higher computing power to find new block. After finding new block or finished mining the system switch PoS. new finding block contain header or address of miner's reward. This header detail is used to validate the new blockchain.

Delegated Proof of Stake (DPoS)

To generate the blocks according to the stake, miners similar to POS get their priority. The main difference between DPoS and PoS is DPoS is the representative democratic while PoS is the direct democratic. To validate and generate the block, stack holder choose their own delegates. The block could be quickly confirmed and the transaction make quickly confirmed by using the significantly a few nodes to formalize a block. In the meantime, the network parameter could be tuned such as interval of block and size of block. Since a delegate could be easily voted out, so users should not worry about the delegates who are dishonest. DPOS is the Bitshares's backbone and it has been already implemented (Bach and Mihaljević, 2018).

Ripple

It is an algorithm of consequence which utilizes the subnetworks of collectively trusted inside the larger network. Nodes are divided in a network into two types: first one is server that participating in the process of consequence and second one is client that is used only for fund transferring. In the network, nodes of Practical Byzantine Fault Tolerance (PBFT) have to ask each node in the contrast and for query every server of ripple has a unique node list (UNL) that is important for server. When into the ledger whether determinant put a transaction than in UNL the server would query nodes. If the agreements that received have reached till 80% than into the ledger transaction would be packed. The ledger for a node will remain correct as far as the faulty nodes percentage is less than 20% in UNL (Nguyen, 2018).

Tendermint

It is an algorithm of byzantine. A new block is specified in a round. In this round to broadcast, a block that is unconfirmed, a proposer would be selected. So, for selection of proposer each node required to be known. It could be separated into three steps:

- Prevote step: For the block which is proposed, validators select whether a prevote to be broadcasted.
- Precommit step: If on the block which is proposed the node has received prevotes more than 2/3 then to the block it transmits a precommit and nodes enter the step of commit if it received precommit over the 2/3.
- Commit Step: The block validated by the node and for that block it broadcast the commit. If node has got commit 2/3 then the block is accepted. This process is quite a similar to PBFT, but to become validators the nodes of Tendermint have to lock there coins. A validator would be penalizing if it is detected once to be dishonest (Nguyen, 2018).

2.11   CONSENSUS ALGORITHM COMPARISION

Different algorithms of consensus have different disadvantage and advantage. Comparison between dissimilar algorithm of consensus is shown in Table 2.1 and we use the following properties given by Vukolic:

- **Node Identity Management:** The identity of every miner requires to know by PBFT in order so that in each round a primary is selected while Tendermint requires to know validators in every round in order to choose a proposer. Network could be joined freely by nodes for PoS, Ripple, DPoS, and PoW.
- **Energy Saving:** In PoW to gain the value of target, the header of block is hashed by miner continuously. The electricity amount is needed to process as a result which has scale of immense. Miners have to hash header of block for PoS and DPoS to find the value of target but as the space of search is designed limited so the work has been trimmed greatly. In the process of consensus, there is no one mining as for Ripple, Tendermint and PBFT. Thus, it is used to save the energy largely.

- **Tolerated Power of the Adversary:** To gain network control to one as threshold, 51% of power of hash is generally regarded. But the strategy of selfish mining in system of PoW could support miners by 25% of the power of hashing to acquire revenue more. To handle nodes which are faulty up to 1/3, Tendermint and PBFT are designed. If the nodes that are faulty is less than 20% in UNL than to maintain the correctness ripple has to proved.
- **Example:** Peercoin is a newly peer-to-peer cryptocurrency of PoS while bitcoin is based on POW. Farther to gain consensus Fabric of Hyperledger utilizes PBFT. Bitshares which is a smart platform of contract that adopt the DPoS like their algorithm of consensus. Protocol of ripple is implemented by ripple while the protocol of Tendermint is invented by Tendermint.

  Tendermint and PBFT are permitted protocol. Identities of node are expected for the complete network to be known, so in mode of commercial they might be used and not used in public mode. For blockchain of public, POS and POW are suitable. Blockchain of private or consortium might have orientation for Ripple, DPOS, PBFT, and Tendermint (Nguyen, 2018).

2.11.1   ADVANCE ON CONSENSUS ALGORITHM

An effective algorithm of consensus means convenience, safety, and efficiency. Current usual algorithms of consensus still have more shortages. Algorithm of new consensus is formulated with aiming to solve some particular blockchain problem. The peer consensus primary idea is to uncouple creation of block and confirmation of transaction so that speed of consensus can be increased significantly. Besides, a new method of consensus is proposed by Kraft to assure that block is generated in a comparative speed that is stable. It means the generation rate of high block compromises the security of Bitcoin. So to solve this type of problem a new rule is proposed that is Greedy Heaviest Observed Sub-Tree (GHOST) chain selection rule. Rather of the scheme of longest branch, GHOST weights of the subdivisions and miners could select the one of the better to follow (Nguyen, 2018).

2.12   CHALLENGES AND RECENT ADVANCES

Scalability

The blockchain gets heavy as day by day transaction amount increases. Currently, the storage of bitcoin and blockchain has exceeded 100 GB. For formalizing the transaction, all the transactions have to be stored. Besides, blockchain cannot satisfy the processing requirement of millions of the transaction in the fashion of real time because of the interval of time which is used to get a new block and original restriction of size of block, so the blockchain of bitcoin can process only seven transactions nearly per second. In the meantime, many transactions that are small might be holdup because of the block capacity is very small since miners choose those transactions which have a high fee of transaction. Although, the size of large block would slow down the speed of propagation and guide to branches of blockchain. So, the problem of scalability is quite tough. Here to address the problem of scalability of blockchain, a number of efforts are proposed that could be categorized into two types:

- **Storage Optimization of Blockchain:** A scheme of novel crypto-currency was proposed to solve the problem of bulky blockchain. In this new scheme by the network, old records of transactions are removed and to store the balance of all address that is non-empty a database is issued named account tree. So in this scheme to check the valid or invalid transaction, nodes don't require to hold all transactions. Besides clients that are lightweight could also help to solve this type of problem. A scheme of novel was proposed named is VerSum to offer the other way which allow the clients that are lightweight to exist. VerSum allow the clients that are lightweight for expensive computation of outsource over the big inputs. By comparing the results from various servers, it assures that the result of computation is correct.

- **Redesigning Blockchain:** Bitcoin Next Generation (NG) was proposed by Eyal et al. The aim of Bitcoin NG is to uncouple the block of conventional into two parts: first one is key block that is used to elect leader and second one is micro block which is used to store the transaction. To become the leader, miners are contended. Until a new leader appears, that leader would be creditworthy for generation of micro block. Bitcoin NG also expended the strategy of chain that is longest where the micro blocks don't carry any weight and only key blocks are counted. In such a way, the trade-off has been addressed which is between the security of network and size of block and it redesigned the blockchain (Zheng, 2018).

Privacy Leakage

The chain of block is believed to be very secure as user do only the transaction with the address that is generated instead of the real identity. In case of leakage of information users could generate various addresses. Although, blockchain can't assure the privacy of transaction since the values of balance and whole transactions are visible publicly for every public key. A Bitcoin transaction of user can be connected with the information of disclose user. Furthermore, a method is presented to connect the pseudonyms of user with the IP address yet when the users are backside of firewalls or network address translation (NAT).

Set of nodes which is connected with client is used to uniquely identify every client. Although this set of node can be used to find the transaction origin and to be learned. To improve blockchain anonymity, various methods are proposed that could be categorized roughly into two types:

- **Mixing:** Addresses of user in blockchain are anonyms. But it is yet possible to connect address with the real identity of user as various users do transactions with the frequently of same address. Service of mixing is that type of service which gives namelessness through the fund transferring from various input addresses to various output addresses. Such as user Aline with A address wish to transmit fund to Bob with B address. If Aline directly does the transaction with A input address and B output address, then relationship among Aline and Bob might be disclosed. So Aline could transfer funds to the trusted Carol who is the mediator. Than mediator send funds to Bob with the help of various inputs like c1, c2, and c3 and various outputs like d1, d2, d3, and B etc. Address of Bob B is also considered in output addresses. Thus, it gets harder to disclose relationships among Aline and Bob. Although, the mediator could be dishonorable and disclose the private information of Aline and Bob purposely. That is also possible for Carol that he can transfer the fund of Aline to her own address rather than address of Bob. So a simple method is provided by Mixcoin to avoid the dishonorable behaviors. The mediator encrypt the requirements of user in which include date of fund transfer and amount of fund with its private key. If mediator didn't send the money then anyone could verified that mediator cheated with them. Although, theft is discovered but yet not prevented. To prevent the theft, addresses of output is shuffled by CoinJoin which is dependent on central server of mixing and it is inspired from CoinShuffle and CoinJoin which is used in decryption of mixnets for the shuffling of address (Zheng, 2018).

- **Anonymous:** A proof of zero knowledge is used in Zerocoin. Miners validate the coins which belong with a list of valid coins but the transaction do not validate with digital signature. Origin of payment is different from the transaction to graph analyzes of prevent transaction. But it even discloses the destination of payment and amounts. To solve this type of problem, a Zerocash method was proposed. Zero-knowledge summary and arguments that is non-interactive of knowledge is purchased in Zerocash. The amount of transaction and the coin value are hidden which is held through user.

Selfish Mining: The blockchain to conspiring attack of selfish miners is susceptible. Mostly, it converts the nodes which is with over the 51% power of computing that could inverse the transaction which happened and also inverse the chain of block. Although, the nodes which have power less than 51% are even dangerous. If even some portion of power of hashing is used for cheat then network becomes vulnerable. In the strategy of selfish mining, the selfish miners without the broadcasting keep their blocks of mined and if some needs are satisfied then the private branch would be disclosed for public. As the current public chain is smaller than the private branch, so through whole miners it would be admitted. The honest miners are atrophy there resources on a branch which is useless before the publication of private blockchain while without the competitors, miners who are selfish are mining there private chain. So the miners who are selfish tend to generate greater revenue. Miners who are intellectual would be attracted to link with selfish pool. Selfish could quickly exceed with 51% power. To show the unsecure blockchain, various other attacks which are based on selfish mining have been proposed. In obstinate mining, the miners expand its profit through the composing of nontrivially attacks of mining with the network level attack of eclipse. The lead stubbornness is the strategy of one of the stubborn in which still block is mined by miner even if private chain is left over behind. Still in some time, it gains 13% result in equivalence with the nonlead stubborn similitude. Strategies of selfish mining are shown by Sapirshtein in which they show that for small miners it is very profitable and they gain more money comparatively with simple selfish mining. But earning is comparatively small. Moreover, it shows that even if the resources of computational is with less than 25% than the attacker still can earn from the selfish mining. To solve the problem of selfish mining, a novel approach is proposed for honest miners that choose the branch which they want to follow. Honest miners would choose with random timestamps and beacons more fresh blocks. Although for forgeable timestamp it has become vulnerable. Zeroblock establish in the scheme that is simple. Every block should be accepted and generated within a maximal interval of time through the network. Selfish miners inside the Zeroblock cannot gain more than its reward which is expected (Zheng, 2018).

2.13 CONCLUSION

Blockchain technology runs the bitcoin digital cryptocurrency. It supports decentralized environment persistency, anonymity, and auditability for transactions of records. Each transaction is recorded into the public ledger, which is visible by everyone in network. The motive of blockchain is that it gives user data security, transparency of data, anonymity, and privacy. In this chapter, we first give overview of blockchain technology that include its functional characteristics and architecture of blocks. Then we discuss about the classification of blockchain where we discuss about public, private, and hybrid blockchain, which depends on the blockchain application and also discuss that how blockchain technology works? While moving forward this link, we have discussed the algorithm for various consensus algorithms and also compared the various approaches with each other. However, these technologies have some technical issues and challenges that need to be addressed which are also described in this chapter.

KEYWORDS

- **blockchain**
- **bitcoin**
- **cryptocurrency**
- **cryptography**
- **digital signature**

REFERENCES

Agrawal, R. Digital Signature from Blockchain Context, 2018.

Atlam, H. F. Technical Aspects of Blockchain and IoT, 2018.

Bach, L. M.; Mihaljević, B.; Žagar, M. Comparative Analysis of Blockchain Consensus Algorithms Rochester Institute of Technology, Croatia, MIPRO 2018/SP.

Blockchain Technology Primer IAB Tech Lab Version 1.0 | 2018.

Blockchain Technical Series Katalyse.io Blockchain Consensus—What You Should Know, 2018.

Kehrli, J. Blockchain Explained, 2016.

Khatwani, S. What Are Private Blockchains & How Are They Different from Public Blockchains?, 2018.

LegalTech News A Brief History of Blockchain 2018.

Lisk   Academy   https://lisk.io/academy/blockchain-basics/how-does-blockchain-work/ digital-signatures

Marr, B. A Very Brief History of Blockchain Technology Everyone Should Read, 2017.

Naidu, G.; Mishra, R. Blockchain Technology Architecture and Key Characteristics **2018**, *4* (4).

Nguyen, G. T.; Kim, K. A Survey about Consensus Algorithms Used in Blockchain. *J. Inf. Process Syst.* **2018**, *14*, (1), 101–128.

Pantas and Ting Sutardja. BlockChain Technology Beyond Bitcoin. Center for Entrepreneurship & Technology Berkeley Engineering.

Puthal, D.; Mohanty, S. P.; Malik, N. S.; Kougianos, E. Everything You Wanted to Know About the Blockchain: Its Promise, Components, Processes, and Problems, 2018.

Sultan, K.; Ruhi, U.; Lakhani, R. Conceptualizing Blockchains: Characteristics & Applications 11th IADIS International Conference Information Systems 2018.

Vujicic, D.; Sinisa, R. Blockchain Technology, Bitcoin, and Ethereum: A Brief Overview, 2018.

Zheng, Z.; Xie, S.; Dai, H.; Chen, X.; Wang H. An Overview of Blockchain Technology: Architecture, Consensus, and Future Trends. *IEEE 6th Int. Congress Big Data.* **2017**.

Zheng, Z.; Xie, S.; Dai, H.; Chen, X.; Wang, H. Blockchain Challenges and Opportunities: A Survey. *Int. J. Web Grid Serv.* **2018**, *14*, 4.

Zilavy, T. What Is A Hybrid Blockchain And Why You Need to Know About It?, 2018.

CHAPTER 3

Blockchain, Bitcoin, and the Internet of Things: Overview

JYOTIR MOY CHATTERJEE,[1*] P. SRINIVAS KUMAR,[2] ABHISHEK KUMAR,[3] and B. BALAMURUGAN[4]

[1]*Department of IT, LBEF(APUTI), Kathmandu, Nepal*

[2]*Department of CSE, SSUTMS, MP, India*

[3]*Computer Science & Engineering Department, Chitkara University Institute of Engineering and Technology, Chitkara University, Himachal Pradesh, India*

[4]*Department of CSE, Galgotias University, Greater Noida, UP, India*

Corresponding author. E-mail: jyotirchatterjee@gmail.com

ABSTRACT

The "Big Four" advancements to come are starting to compel themselves to the fore as associations fiddle with any semblance of artificial intelligence (AI), blockchain (BC), internet of things (IoT), and big data (BD). Be that as it may, as these four springs from their embryonic stages, issues, and concerns are being paid attention to about their development. BC has impeded in its reception and development; however, its application is wide and diverse. In this way, it can profit, and help advantage, any semblance of AI and IoT. Be that as it may, BC can likewise have a critical task to carry out in aiding along with IoT. The development of IoT is as yet moving along, yet it also has hindered on the grounds that many have understood that acing this system of smart 'things' is far harder than they could have envisioned. Issues of security and courses of events of execution have caused the publicity around IoT to cool off. In a comparable vein, this has additionally occurred in BC. It is on the grounds that IoT and BC end up in comparative spots with regards to the selection that their mix might have the capacity to enable each other to out and comprehend a portion of the noteworthy worries that have hounded them exclusively. BC is no more in its early stages, yet it is the latest till now. Comparative proclamations can be done regarding the IoT. The buzz around BC utilization in IoT, be that as it may, is undeniably later. The association of the two skirts on the untested—and right now, the unapplied. In the IoT situation, the square chain and, when all is said in done, peer-to-peer methodologies could assume an essential job in the advancement of decentralized and data-serious applications running on billions of gadgets, saving the security of the clients. In this chapter, we will try to provide a detailed overview about this linkage between BC, Bitcoin, and IoT. We will focus on how IoT problems can be solved using the help of BC technology and vice versa.

3.1 INTRODUCTION

3.1.1 BLOCKCHAIN

BC concept was brought into existence in the year of 2009 by an anonymous person Nakamoto (2008). First this mechanism was used in bitcoin technology which is a digital currency mechanism for security purpose while transaction. Here, the term block is used in the concern of data storage. Every block holds a value which points to the next value in a form of linked-list (Chatterjee. 2018). This link format helps for data security, uniqueness of data, correctness, and efficiency of data. The block structure comprises of header section as well as the body section. The block header has block size, version, and previous block link. Block body has transaction as well as "The Merkle tree." "The Merkle tree" has a leaf node which maintains internal nodes as well as verification blocks. These enable the security to the blocks (Chatterjee, 2018). The blockchain can verify the results firstly between the nodes during the transmission process. The bitcoin the nodes are collected into blocks. The BC depends on two cryptographic mechanisms they secured digital manner. It is a mechanism in explaining the security of a message which was in digital form. It can also use by providing integrity and authentication of data and nonrepudiation. This data verifying process can be done by anyone (Council Post, 2019). A cryptographic hashing function is an operation which performs hashing values on given data. The function is defined, such that, the same type of input results to the same output, by obeying features like preimage refusal second preimage refusal, along with collision handling capacity. BC stands for Bitcoin, a SHA-256 hashing mechanism was used (Das, 2015).

The BC model depends on cryptographically protected; unchanged shared ledger mechanism along consensus can extend internet of things (IoT) frameworks by automatically resource extraction as well as innate security by providing:

- A distributed system of record for sharing data across a network of key stakeholders.
- Embedded business terms for automating interactions between nodes in the system.
- Security based on hashing mechanisms, verifying the identity as well as authentication.
- Consensus, as well as agreement mechanisms, for detecting bad actors and mitigating threats.

With such features, a BC-enabled IoT deployment could improve overall system health and integrity by allowing devices to register and validate themselves against the network. Business logic could execute automatically via smart contracts. And without any central system in attacking, threats like Denial of Service could be inherently deterred at different layers in the architecture. Applying a BC model in an IoT network could solve a host of real-world digital business issues, including

- Analytical model tracking: Allows the system to record metadata and results about logic executed at the edge of the network for the purposes of regulatory compliance and create an immutable history of why certain "decisions" were made during IoT processing.
- Secure software updates: The ability to publish software updates as a URL on the BC, along with a cryptographic hash of the update which can be validated by BC-connected IoT devices during the process.
- Payments and micropayments: Automated payments to business network participants based on sensor data (indicating, for example, service completion or product delivery), as well as micropayments between devices themselves in certain networks for functions and capabilities—all without human involvement (Gross, 2015).

3.1.2   IOT

The concept of IoT became the most significant technology in this century. This is the mechanism where the computers are connected to the devices with the help of embedded programming. Here the things are interconnected to the systems and servers for task completion automatically with more secured and robustness that type of knowledge which gives efficiency as well as provides advanced services in more range of domains which includes health monitoring as well as smart city service. After all, the increase in invisible as well as collection, the procedure of dissemination of information in the middle of users' private live gives (Badev and Chen, 2014). On the other end, this information can give a certain amount of practical, as well as customized assistants, which gives utilization to users. On the other end, here the information can be used to procedurally construct virtual data regarding our day-to-day activities (Amoozadeh et al., 2015).

The basic problems of IoT are due to the scarcity of basic privacy problems over IoT the data items on the market. Largely we found privacy is only a major issue in major IOT devices (Bider and Baushke, 2012) to vehicles (Blockchain - Bitcoin Wiki, 2019). Several security and confidentiality problems that include: Lack of control, users accessing from different resources, multiple attack surfaces, context-aware, as well as scale. Privacy is a challenging task in IoT Technology. In Badev and Chen (2014), a distributed mechanism access control mechanism was proposed for controlling access to data. Whatever, the methodology introduces excess of latency time along with overheads that can potentially compromise user privacy. Authors in Bloomberg (2019) use IPsec and TLS in providing user credentials as well as confidentiality, but these procedures are practically more cost along that may be different to many source-providing IoT devices. A confidentiality mechanism was introduced (Bloomberg, 2018) in which it calculates the problem of hiding data to others. Yet, in more cases, the advantage of IoT service regarding the problem of security. There is a scarcity of data exchange in a secure manner for IoT information without obeying the security concerns for users. The following are the present challenges in the field of IoT (Bider and Baushke, 2015):

- Decentralization: The lack of control over the centralized server for communicating client from one server to the other server. If the communication disturbs the sending and receiving data from that node stops.

- Anonymity: The extraction of anonymity is more flexible for more IoT users, here it will be confidential.
- Security: BC reorganizes IoT devices for securing. The IoT devices provide service for the homogenous type of data as well as the heterogeneous type of data. Most of the security problem is lagging in sharing the data in different resources that include:
- Extraction of the data is the difficult and time taking task while most of the IoT devices are resource restricted.
- Extracting the data in blocks takes more time for data receiving from the server by most IoT devices with low waiting time is needed.
- BC rated very less as the blocks increase over the grid the data collection too takes more time.
- The BC algorithms that create more traffic, that are undefined for specific transmission capacity that limits to IoT equipments. Presently IoT became the major topic, where the processing is done by the combination of hardware and software devices and data is stored in database for storing the data.

Gartner has warned that three-quarters of all IoT projects will take twice as long as planned to implement, and IoT security (or a lack thereof) has been called "a doomsday scenario waiting to unfold." It turns out that IoT is hard! Tackling this complexity is core to capturing the promised benefits of IoT. In today's increasingly digitalized world, the ability to make sensors, devices and computational "things" perform tasks and functions for us is becoming a necessity. Human beings just cannot manage the explosion of data and "interconnectedness" on their own, but we also do not need bad bots running amok. Establishing helpful IoT systems that run securely, efficiently, and independently has proved incredibly difficult. BC shows promise for easing that burden (Blockchain - Bitcoin Wiki, 2015).

3.1.3 IOT CHALLENGES

Cybercrime that exploits IoT devices and networks shows no sign of abatement. Thus, security, privacy, and identity verification remain foundational concerns in IoT deployments. A huge amount of IoT information must be collected, transferred, as well as delivered in a secure fashion among valid stakeholders and processing now occurs at various layers within architecture to trigger decisions at the right point. Misbehaving sources must also be detected and resolved, as device-related threats may include (Chatterjee et al., 2018):

- Physical device attacks (unauthorized device control).
- Software attacks (malware such as viruses or worms).
- Network attacks (denial of service assaults, wireless vulnerability exploits).
- Encryption attacks (brute-force password cracking, "Man in the Middle" attacks).

The fact that there are so many types of threats and so many unsecured or poorly secured devices (and no standard authority for connectivity requirements) further confuses matters. Add to this the growing cadre of IoT communication protocols and you have an extremely complex problem that is difficult to solve at scale (Chatterjee, 2018).

3.1.4   IOT TODAY

In spite of the Byzantine complexity and myriad threats involved, IoT adoption continues to accelerate with analysts forecasting nearly $15 trillion in aggregate IoT investment by 2025. Why? The 2017/18 Vodafone IoT Barometer survey of 1278 enterprise and public sector executives from 13 countries holds a clue: 74% of respondents who have adopted IoT technologies claim digital transformation is impossible without it. Basically, it does not matter if it is hard, it has to be done if you want to do business in the modern world (Chatterjee et al., 2018).

Using the Industrial Internet Consortium's reference architecture as an example, a basic business IoT system will have enterprise, platform, and "edge" tiers managing the devices themselves in addition to contextual rules, events, business state, multiple protocols, various data formats, supporting analytical models, and storage—all through varied and dynamic networks. Traditional IT systems are not sufficient. Enterprise systems will need to be extended and enhanced for handling the volume, velocity along a variety of data generated by IoT networks as well as the ability to trigger correct decisions against trusted data will need to be enabled at each level in the architecture (Council Post, 2019).

This represents an awful lot of potentially vulnerable technology moving an awful lot of potentially vulnerable data and an awful lot of complexity to manage. Never mind adding the notion that we will soon have devices "paying" each other for capabilities without human involvement, thus adding to the complexity of emerging IoT systems (Das, 2015).

3.1.5 THERE IS NO SUCH THING AS A MAGIC BULLET

In spite of all this potential, applying BC to IoT is not a cure-all. Current performance and scalability limitations are incompatible with many IoT functions. External data must be incorporated via trusted "oracles." A new type of BC platform supporting the volume of devices involved in an IoT deployment is needed, with capabilities that extend beyond today's common models. What will likely emerge is a hybrid or polyglot architecture, with varying frameworks customized to utilize BC differently at IoT's edge, platform, and enterprise layers. But as businesses continue to grapple with core IoT complexity and security problems, it is becoming obvious that BC-based solutions have merit and bring real value to the table. BC is not the answer to everything that ails IoT but it can play a powerful role in solving some serious issues. It would not save IoT, but it might just improve it (Gross et al., 2015).

Despite fog computing along with fraud information over BC, it has different mechanisms in IoT. Ranges by device communication securely to shared information creation along automatic data auction, BC limitations in IoT have advantages. We cannot make all the applications a pilot test is performed for making the tests the standard one (Nakamoto, 2008).

3.1.6 IOT DURABILITY AND BC VOLATILITY

Numerous IoT sensors were intended for choice in the area of a considerable time period. On the opposite side is BC, where general market condition still unpredictable. The successful IoT devices need dependability, versatility, and continuance. Expecting we settle the fundamental task among two verticals along gave us chooses to join BC along IoT—we keep running onto other issues. The prior BCS, as well as related host organizations, were unpredictable, attached for a hypothesis, and untested (Jha et al., 2019).

The arrangement of constructing by own that can be expensive along restrictively tedious. In terms of cost, presently appears as though an opportunity to make reference to open-source, BC-dependent venture from IBM. Hyperledger is easy as well as with user-customized options to begin with. It's likely extraordinary compared with other approaches to assemble a BC application all alone. Indeed, even with Hyper ledger, building BC is not a simple undertaking. You need to discover people who are specialists in BC along IoT with the goal that it can settle on wise choices regarding the trap among the two.

3.1.7   A GAP IN THE CHAIN: HUMAN ERROR AND SECURITY ISSUES

The secondary problem is privacy. The IoT has "Poor data guarantee." Since BC is to give a solution for all of the security threats. In any case, a BC is just as secure as its data encryption. To date, it is individual composing code along specialists interact once in a while. By capital market drifting mechanism overall $2.6 billion, it is anything but difficult to expect the IoT is error-free, security gaps, along with insane code (Nakamoto, 2008).

The IoT undertaking has hashed everywhere throughout the web. Particle has done one task that ought to never do in PC security. They have done their own cryptanalysis, that opened up to the lot for impact assaults which allow individuals to take or give away IoT to individuals unlawfully. Furthermore, the framework depends on a ternary as opposed to parallel framework that backs off the code as well as makes typical cryptographic capacities hard to utilize. On the off chance that you cannot confide in the ideal example of BC and IoT, who would you be able to trust?

3.2   FUTURE BC APPLICATIONS IN IOT

While BC certainly has some obstacles to overcome, the truth is that the future of BC applications in IoT is exceptionally bright. IDC estimates that by 2019 as much as 20% of IoT deployments will include BC technology. And we need answers to ensure that those "20% of deployments" are as successful, efficient, and secure as possible (Garg, 2018).

Since the age and examination of data is so fundamental to the IoT, thought must be given to securing data for an incredible duration cycle. Overseeing data at all dimensions is not an easy task on the grounds that data will stream crosswise over numerous regulatory limits with various arrangements and expectations.

Given the different innovative and physical parts that really make up an IoT ecosystem, it is a great idea to think about the IoT as a system-of-systems. The architecting of these systems that give business incentives to associations will often be a perplexing endeavor, as big business engineers work to configuration coordinated arrangements that incorporate edge gadgets, applications, transports, protocols, and examination abilities that make up a completely working IoT system. International Data Corporation (IDC) gauges that 90% of associations that actualize the IoT will endure an IoT-based break of back-end IT systems constantly 2017 (Garg et al., 2017).

3.3   CHALLENGES TO SECURE IOT DEPLOYMENTS

Despite the job one's business has inside the IoT community—device production, arrangement provider, cloud provider, systems integrator—you have to realize how to get the best profit by this new innovation that offers such exceptionally various and quickly evolving chances. Dealing with the gigantic volume of existing and anticipated information is overwhelming. Dealing with the unavoidable complexities of interfacing with an apparently boundless rundown of gadgets is confused. The current security advances will assume a job in alleviating IoT hazards; however, they are insufficient (How to Secure the Internet of Things (IoT) with Blockchain, 2019).

3.3.1   DEALING WITH THE CHALLENGES AND THREATS

Gartner anticipated that over 20% of organizations will convey security answers for ensuring their IoT gadgets and services by 2017, IoT gadgets and services will extend the surface territory for digital assaults on organizations, by transforming physical items that used to be disconnected into online resources communicating with big business systems. Organizations should tailor security to each IoT sending as indicated by the extraordinary abilities of the gadgets included and the dangers related to the systems associated with those gadgets.

Creating resolution for the IoT requires uncommon joint effort, coordination, and availability for each piece in the framework and all through the framework all in all. It's conceivable, yet it tends to be costly, tedious, and troublesome, except if a new line of reasoning and a new way to deal with IoT security developed far from the current centralized model (Das, 2015).

3.3.2   THE PROBLEM WITH THE CURRENT CENTRALIZED MODEL

The current IoT ecosystems depend on centralized, expedited correspondence models also called the server/client system. All gadgets are recognized, confirmed and associated through cloud servers that sport enormous handling and storage limits. Association between gadgets should solely experience the web, regardless of whether they happen to be a couple of feet separated. Existing IoT arrangements are costly a direct result of the high framework and support cost related to centralized clouds, huge server cultivates, and networking hardware. The sheer measure of correspondences that should be dealt with when IoT gadgets develop to the several billion will build those expenses generously (Blockchain - Bitcoin Wiki, 2019).

Regardless of whether the extraordinary temperate and designing difficulties are survived, cloud servers will remain a bottleneck and purpose of disappointment that can upset the whole system. This is particularly essential as increasingly basic errands. Also, the assorted variety of proprietorship among gadgets and their supporting cloud foundation makes machine-to-machine (M2M) interchanges troublesome (Badev and Chen, 2014)

3.3.3   DECENTRALIZING IOT NETWORKS

A decentralized way in dealing with IoT systems says a large number of information. Improving peer-to-peer communication mechanism model to process among multiple gadgets that will fundamentally less than expense related with by introducing as well as updating an expansive incorporated master node along with appropriate results.

In any case, setting up distributed networking will show its own arrangement of difficulties among them. What's more, as we as an all know, IoT security is significantly more. The proposed arrangement should keep up the protection along with privacy in tremendous IoT systems as well as provide some type of approval along with accord in exchanging to counteract satirizing along with robbery (Chatterjee et al., 2018).

To play out elements for conventional IoT arrangements with no unified control, any decentralized methodology must help three principal capacities:

- Point to point communication
- Globally file sharing
- Automatic device

3.4   THE BC APPROACH

BC is a "distributed record" innovation, that supports Bitcoin, has risen as an object of exceptional for business. BC innovation offers a method for account exchanges or any advanced cooperation in a procedure which was intended to be safe, very impervious for blackouts and proficient; all things considered, it conveys the likelihood of disturbing ventures and empowering new plans of action. The innovation is youthful and changing quickly; far-reaching commercialization is as yet a couple of years off. Regardless, to dodge troublesome shocks or botched chances, strategists, organizers cross-wise over enterprises, and business capacities should pay notice now and start to examine the uses of the innovation (Council Post, 2019).

BC is a database that keeps up a consistently developing arrangement of information. It is conveyed naturally, implying that there is no PC holds the overall list. Or maybe, partaking hubs having duplicate of block. It is additionally consistently developing—information was just added to the block.

A BC contains two assortments in components:

- Transactions were the activities made by members on the framework.
- Block of record exchanges as well as ensure the right succession along have not been merged up.

The huge preferred standpoint of BC is that it's open source. Everybody taking an interest can see the squares and the exchange's data among them. This does not mean that everyone can view the genuine substance for your exchange, in any case; which is secured by your secret key. A BC is a local server, so there is no expert which can support exchanges. That implies large quantity of trustable required since every one of the members on the system needs to achieve an accord to feedback exchanging. Above all, it is safe. The database must store past records and cannot be modified (Das, 2015).

When somebody needs to exchange the block, every one of the members in the system should approve it. This mechanism is done by applying a calculation to the exchange to check its legitimacy. A lot of endorsed exchanges are then packaged in blocks, which gets sent to everyone must accept from the local devices. Then, they approve the new chain. Every progressive block will have a hash, which is one of a kind unique mark, of the past block. Two main classifications of BC:

- In an open BC, everybody can peruse or compose information. Some open BCs limit the entrance to simply perusing or composing.
- In a private BC, every one of the members is known and trusted. This is valuable when the BC is utilized between organizations that have a place with the equivalent legitimate substance.

3.4.1   THE BC AND IOT

BC innovation is the missing connection to settle versatility, protection, and dependability problems in the IoT. BC advances could be the projectile required by the IoT business. This decentralized methodology would kill single purposes of disappointment, making a stronger biological community for gadgets to keep running on (Gross et al., 2015).

Records were carefully designed and cannot be controlled by invalid users or restricted users. Since it does not present in any area and man-in-the-middle assaults cannot be organized that there is no single string of correspondence that can be caught. BC makes trustless, shared informing conceivable, and has officially demonstrated its value in the realm of money related administrations through digital forms of money, for example, Bitcoin, giving ensured distributed installment administrations without the requirement for outsider merchants. The decentralized, independent, along trustless abilities of BC, make it a part to end up by a crucial term of IoT. In an IoT arrangement, the BC can keep an unchanging record of the historical backdrop of brilliant gadgets. This element empowers the self-governing working of brilliant gadgets without the requirement for brought together specialists. To empower message trades, gadgets will use shrewd contracts which at that point display the attention between the two gatherings (Ho et al., 2016).

In this situation, the sensor from a remote place, discussing specifically with the water system framework so as to discipline the stream of water-dependent on terms identified on yields. So also, brilliant gadgets in an oil stage can trade information to alter working dependent on climate conditions. Utilizing the BC will empower genuine self-governing gadgets that can trade information, or even execute exchanges, without the need of a concentrated representative. This kind of self-rule is conceivable in light of the fact that the hubs in the BC system will confirm the legitimacy of the exchange without depending on a brought together expert (Nakamoto, 2008).

In this situation, we can imagine keen gadgets in an assembling which can put the requirement for fixing a portion for its parts without any of the requirement. The perfect method is to keep all the required clients and servers connected and it should be in a centralized manner it means that all the devices should connect each other by communicating among them and send the result to the server and can query the result from the client for authorized persons. This mechanism is defined to arise the number of compliances, as well as administrative needy, of mechanical IoT applications without the need to depend on a unified model (Ho et al., 2016).

Nakamoto (2008) gives understanding of the elements that accompany the development of IoT in the furnishings and kitchen producing industry. Skarmeta et al. (2014) tried to give a sensible increasingly significant comprehension about the IoT in BD structure close by its distinctive issues and difficulties and focused on giving possible arrangements by ML system. Ukil et al. (2014)have seen the utilization of IoT for building up a framework dependent on IoT which will enable us to check the status of Gas Knob and could spare us from the perilous threat brought about by the superfluous spillage of gas. Chatterjee et al. (2008) audit the most recent advances on IoT with IoC from a class survey of distributed articles from 2009 to 2017.

3.5 BITCOIN AND BLOCKCHAIN

Bitcoin and blockchain are not similar but they are firmly related. At the point when Bitcoin was announced as open-source code, blockchain technology got a good reputation for its extraordinary features. What's more, since Bitcoin was the main use of blockchain, individuals frequently used "Bitcoin" to make the blockchain. That's the means by which the misconception began. Blockchain innovation has since been extrapolated for use in different areas, yet there is still some complexity and issues (What is the bitcoin blockchain? 2019).

3.5.1 HOW BITCOIN AS WELL AS BLOCKCHAIN BOTH VARIES EACH OTHER?

Bitcoin was a type of digital currency firstly designed by Nakamoto (2008) and it is named as "digital currency," it was linked with the cash control databases and rearrange online exchanges by disposing of outsider installment handling in between them. Obviously, achieving this required goal is like achieving something like cashless transactions. There must be a security mechanism to make exchanges with the cryptographic currency.

Bitcoin transfers were put away as well as transferred by using a dataset over a shared system which has open access. The blockchain supports development which keeps the Bitcoin transferred information in the particular database where no one has privileges to access the data or they cannot be edit or removes data.

3.5.2   WORKING MECHANISM OF BITCOIN BLOCKCHAIN?

The Bitcoin blockchain is the simplest for storing data to a record even bit coin transfers records. The record is circulated over a network, that is, without a main authorized function, organize members must concede to the exchange of the data legally. The present understanding, which is treated as "agreement," which is accepted by a mechanism known as "mining."

Later some person uses Bitcoin, mineworkers participate in complex, asset peer–peer network to check the validation of the user before transferring. Through mining, "proof of work (PoW)" which faces some amount of conditions are made. The verification of the task is bitcoin which is to transfer without confirmed by others. For viewing the status of transferred data exchange the users should have a proof that they are valid users or not. They should have authentication and authorization permissions. The study "What is the bitcoin blockchain? (2019)" provided a state-of-the-art survey over Bitcoin-related technologies and sum up various challenges.

3.5.3   HOW IS BLOCKCHAIN USED FOR BUSINESS PURPOSE?

This process for Bitcoin was developed especially for a digital currency which is the main reason to use blockchain for business purposes. In many of the business industry, they are creating blockchain accounts for one company and they are linking for one company to another company by point–point network mechanism. Here only authorized users can log in and transfer data, the data cannot be transferred or modified.

3.5.4   ASSETS OVER CRYPTOCURRENCY

There is a great for bitcoin when it is first introduced. First, they developed a token mechanism system where the token is generated for the user if the user wants to transfer the data. We cannot say something regarding this concept; however, we can define as blockchain is a mechanism that has benefits then paperless money. Improper resources, for example, land and sustenance products, and in an average of different sources, such as the public sector, private sector, cooperate sector and Multinational Companies using this concept. In particular cases, Everledger is using this concept to secure their data and sensitive information like bank accounts information and all. Presently, more than 1 million users are using blockchain.

3.5.5   IDENTITY OVER ANONYMITY

Bitcoin gets hacks more often because of its security. Anybody can take an account of Bitcoin and see whether the information is an arrangement of numbers. The organizations have KYC (know your customer) and AML (anti-money laundering) are the algorithms having certain principles for identifying several information, like user login information, login location, IP address, etc., having in an advantage guardianship mechanism like the one happening produced various gatherings which include funding organizations, overseers, directors, and venture counsels. The users must have to know their own identity and credentials for data transmission and all, only the access privileged persons have a scope to perform all the operations.

3.5.6   SELECTIVE ENDORSEMENT OVER POW

Agreement over blockchain in trade is not possible by mining. Whatever the process is treated as "particular support." It is linked up with the process of business which checks the process action on a daily basis. On off chance the transfer of amount to another bank, to the same bank accounts, the beneficiary's bank that confirms the exchange of the transfer status. The same thing is not in the case of bitcoin; here the overall device has to perform the task to check transferring.

3.5.7   WHY WILL BLOCKCHAIN TRANSFORM THE GLOBAL ECONOMY?

Like how the web changed the world by giving more amount of data for accessing, blockchain is ready to change how individuals work together by offering trust. By plan, anything recorded on a blockchain cannot be modified, and there are records of where every advantage has been. Along these lines, while members in a business system probably would not have the capacity to identify in one another, they can trust the blockchain. The advantages of blockchain for business are various, which includes discovering data, settling the debate and confirming exchanges, diminished expenses, and mitigated hazard (Blockchain - Bitcoin Wiki, 2019).

For all the ways blockchain is as of now being utilized in business, there are many mysteries in the blockchain which are not yet solved (How to Secure the Internet of Things (IoT) with Blockchain, 2019).

At the point when Bitcoin broke into public consciousness in 2013, it could not have been the hot topic: Computerized cash being utilized to purchase everything from medications to the bank accounts. Bitcoin's wild value ride—taking off up in late 2017 just to fall relentlessly withdraw this year—was additionally truly energizing, regardless of whether the swings influenced its utilization as ordinary money to appear to be less conceivable. Another part of Bitcoin, that is somewhat less sparkling, has created nearly as much eagerness: open online records. Blockchain—the innovation utilized for checking and recording exchanges that are at the core of Bitcoin—is viewed as having the capacity to reshape the worldwide money related framework and perhaps different businesses. However, five years after Bitcoin's development, blockchain stays, even more, an enticing idea than a device that is really utilized—sort of like Bitcoin (Bider and Baushke 2012).

The aggregate estimation of Bitcoin was about $300 billion in late 2017, remained at $112 billion in by October end. Over the business thought the presentation of Bitcoin prospects decreases by Cboe Global Markets, CME Group Inc. Furthermore, Nasdaq by the end of 2017 would help assemble authenticity; however, exchanging on the two trades was insignificant. Bitcoin's rollercoaster ride was duplicated by a heap of new digital currencies that have raised more than $21.5 billion through introductory coin contributions (Council Post: Is Blockchain The Way to Save IoT? 2019). Their dangerous development drew admonitions from controllers around the world even before programmers stole nearly $500 million worth of an advanced token called NEM from a Japanese digital money trade. By far most of ICOs lost significant incentive in 2018. Blockchain likewise had various difficulties, as a few prominent activities were dropped or put on hold, including an arrangement by Australia's stock trade to begin utilizing the innovation to process value exchanges. Be that as it may, Wal-Mart Stores Inc. reported in September that beginning in 2019, it will require providers of crisp, verdant greens to follow their items utilizing a computerized record created by International Business Machines Corp., which has spent intensely on creating blockchain devices for business (Das, 2015).

The transferring of the money digitally is not new to the users before the bitcoin was not introduced users are using online payments and online transactions. Bitcoin technology developed by Nakamoto (2008) who also solved a problem regarding currency authorization and server control. They also found a problem like they were able to identify the amount users took was not debited from the account and they are taking the amount twice. By this bug, many of the banks got a great loss in an online transaction. Later, they maintained a network of "miners" for performing computational and mathematical operations. It also helps for checking the state of the transaction whether the transaction is successful, failure, waiting state, aborted state, and dead state. Here in the bitcoin if any user makes to join a new user, they will get the reward points to encourage the new users (Hu, 2003).

Bitcoin is based on the idea of "verification of task." For Bitcoin's situation, verification of work is made through the way toward "mining." To mine a bitcoin, a PC must finish a calculation, regarding persons' account money. Basically, experiencing is done by a broad computation in exchange for the printed money. The money merits, however, the position is outside the vendors checks the position of the amount and security concerns of the amount. The blockchain is a centralized innovation. An overall device utilizes the concept of storing and retrieving the data with the help of the retrieval policy from the database. A blockchain is a database with client appropriated approval. Bitcoin was intended to be only digital money, so it utilizes blockchain to record the bitcoin exchange. In the meantime, there was intended to enable designers to make decentralized applications and it utilizes blockchain to run an application programming code. The blockchain is an advanced record that gives a safe method for making and recording an exchange, assertion, and contracts whatever should be recorded and confirmed as having occurred (Badev and Chen 2014).

Bitcoin helps to speed up the transactions without any transaction loss by obeying the features like atomicity, consistency, isolation, and durability. Blockchain provides low cost and secure environments for a transaction.

Bitcoin technology is limited to digital currency and Bitcoin data stress in encrypted form, whereas Blockchain can easily transfer anything from one user account to another and the blockchain is a ledger.

1. Blockchain innovation enables funds and helps users for safe data sending and retrieving.
2. There are many features in the blockchain which are adaptable features like connections, access privileges, and security.
3. The Blockchain applies Proof-of-Work calculation to approve the procedure.
4. Bitcoin exchanges are put away and exchanged by means of a dispersed record on a shared system which has open access.
5. Bitcoin centers on bringing down the efforts on other devices and it is less preferred than the blockchain.
6. When Bitcoin clients need to transfer their Bitcoin's into cash, they can utilize digital currency to give them for money else different Cryptocurrencies.

7. Bitcoin was made as computerized, decentralized cash, upheld by innovation that would guarantee security for its clients by drawing in its client specifically a job generally held by the banks.
8. A Blockchain is a circulated record taking into account clients to keep up permanent, constant records appropriated over a system.
9. Bitcoin, as a result, wiped out the job of the confided in the outsider by doling out that job to anyone who needed to partake far and wide.
10. Blockchains used to follow the stock, the development of illness, or settle installments rapidly.
11. Bitcoin is based on blockchain innovation, bringing about exchanges followed and recorded on a commonly kept up however unchangeable database.
12. What makes blockchains one of a kind is their written history each exchange ever to happen on the system remains completely discernible to its birthplace, making blockchains things are profoundly safe and secure to deceitful exercises like wholesale fraud.
13. The Bitcoin and blockchain are not the same. Notwithstanding, they are firmly related. At the point when bitcoin is discharged as a freely available software for a similar arrangement. What's more, since bitcoin was the principal utilization of blockchain, individuals regularly unintentionally utilized bitcoin and blockchain.

One of the main and important features in bitcoin and blockchain is versatility. Bitcoin is the money that is used to diminish the exchange and the time of the different exchanges. Bitcoin is a disseminated record which empowers in the more secure distributed computing and in a condition. Be that as it may, in light of the fact that this database is circulated over a shared system and is without an imperative expert, organize members must concede to the legitimacy of exchanges before they can be recorded. Blockchain and Bitcoin have regularly been the idea of being one and the equivalent, and one cannot accuse any individual who has a place with this school of thought (Jha et al., 2019; Chatterjee, 2018).

3.5.8   WHAT IS A BLOCK?

A Block alludes to a lot of Bitcoin exchanges from a specific day and age. Squares are "stacked" over one another so that one square relies upon the past. As such, a chain of squares is made, and subsequently, we go to the expression "blockchain" (Blockchain - Bitcoin Wiki, 2019).

Finding and distributing new squares is the thing that Bitcoin mineworkers do to procure bitcoins. At whatever point another square is a communicated, around at regular intervals, a number of bitcoins is gotten by the mineworker who fathomed that square. Bitcoin diggers keep the system secure and this is the manner by which they are compensated. This framework guarantees that all exchanges are substantial and keeps the bitcoin organize secure from misrepresentation.

On the off chance that you have at any point sat tight for another bitcoin exchange to be affirmed, you were sitting tight for another square to distributed containing your exchange. At the point when that occurs, the bitcoin arrangement has regarded your exchange legitimate. Coinbase at present requires three arrange affirmations before the exchange is considered finished, in any case, this number will shift with other Bitcoin administrations (Lucas and Lucas, 2019; Council Post: Is Blockchain The Way to Save IoT? 2019).

3.6 CONCLUSIONS

Blockchain is being tried in a few enterprises. What's more, now and again we see that in those ventures there is a connection among blockchain and IoT inside the explicit business. A standout amongst the most eager thoughts in blockchain is that the innovation could empower not simply individuals and organizations to execute with one another flawlessly, yet in addition machines. Blockchain can fight off security dangers to Internet-empowered gadgets by giving a distributed ledger to their working, subsequently taking out the central node that networks more often than not rely upon for management by the clients, state specialists. Blockchain innovation utilizes cryptographic methods to make an open and decentralized group of information, which can incorporate keeping money exchanges and other functionalities (Hu et al., 2003). The information record can be checked by anybody associated with the exchange and can be followed by means of a protected system. Blockchain has been most noticeably utilized for cryptographic money exchanges including bitcoin. The present article we tried to present the linkage between Blockchain technology with IoT along with the Bitcoin technology. Be that as it may, both bitcoin and BC are firmly related. At the point when Bitcoin was made as open-source code, blockchain was wrapped up together with it in a similar arrangement. What's more, since Bitcoin was the initial application of BC, individuals often unintentionally utilized

"Bitcoin" to mean blockchain. That is the means by which the misconception began. Blockchain innovation has since been extrapolated for use in different enterprises, yet there are still some confusions persist that need to be solved for utilizing the maximum advantage of the technologies (Lucas and Lucas, 2019).

KEYWORDS

- **artificial intelligence**
- **blockchain**
- **internet of things**
- **big data**
- **bitcoin**
- **peer-to-peer network**

REFERENCES

Amoozadeh, M.; Raghuramu, A.; Chuah, C. N.; Ghosal, D.; Zhang, H. M.; Rowe, J.; Levitt, K. Security Vulnerabilities of Connected Vehicle Streams and Their Impact on Cooperative Driving. *IEEE Commun. Mag.* **2015,** *53* (6), 126–132.

Chatterjee, J. M.; Kumar, R.; Khari, M.; Hung, D. T.; Le, D. N. Internet of Things Based System for Smart Kitchen. *Int. J. Eng. Manuf.* **2018,** *8* (4), 29.

Chatterjee, J. IoT with Big Data Framework using Machine Learning Approach. *Int. J. Mach. Learn. Net. Collab. Eng.* **2018,** *2* (2), 78–85. DOI: 10.30991/ijmlnce.2018v02i02.005.

Chatterjee, J. M.; Ghatak, S.; Kumar, R.; Khari, M. BitCoin Exclusively Informational Money: A Valuable Review From 2010 to 2017. *Qual. Quant.* **2018,** *52* (5), 2037–2054.

Council Post: Is Blockchain The Way to Save IoT? 2019. Retrieved from https://www.forbes.com/sites/forbestechcouncil/2018/07/18/is-blockchain-the-way-to-save-iot/

Das, M. L. Privacy and Security Challenges in Internet of Things. In *International Conference on Distributed Computing and Internet Technology*; Springer: Cham; 2015; pp. 33–48.

Badev, A. I.; Chen, M. Bitcoin: Technical Background and Data Analysis, 2014.

Bider, D.; Baushke, M. SHA-2 Data Integrity Verification for the Secure Shell (SSH) Transport Layer Protocol (No. RFC 6668), 2012.

Blockchain - Bitcoin Wiki, 2019. Retrieved from https://en.bitcoin.it/wiki/Block_chain

Blockchain Applications in IoT: Good and Bad | IoT For All, 2019. Retrieved from https://www.iotforall.com/blockchain-applications-in-iot/

Bloomberg - Are You a Robot? 2019. Retrieved from https://www.bloomberg.com/quicktake/bitcoins

Garg, S.; MoyChatterjee, J.; KumarAgrawal, R. Design of a Simple Gas Knob: An Application of IoT. In *2018 International Conference on Research in Intelligent and Computing in Engineering (RICE)*, IEEE, 2018, pp. 1–3.

Gross, H.; Hölbl, M.; Slamanig, D.; Spreitzer, R. Privacy-aware Authentication in the Internet of Things. In *International Conference on Cryptology and Network Security*; Springer: Cham; 2015; pp. 32–39.

Ho, G.; Leung, D.; Mishra, P.; Hosseini, A.; Song, D.; Wagner, D. Smart Locks: Lessons for Securing Commodity Internet of Things Devices. In *Proceedings of the 11th ACM on Asia Conference on Computer and Communications Security*, ACM, 2016, pp. 461–472.

How to Secure the Internet of Things (IoT) with Blockchain, 2019. Retrieved from https://datafloq.com/read/securing-internet-of-things-iot-with-blockchain/2228

Hu, Y. C.; Perrig, A.; Johnson, D. B. Efficient Security Mechanisms for Routing Protocolsa. In *NDSS*, 2003.

Jha, S.; Kumar, R.; Chatterjee, J. M.; Khari, M. Collaborative Handshaking Approaches Between Internet of Computing and Internet of Things Towards a Smart World: A Review From 2009–2017. *Telecommun. Syst.* **2019,** *70* (4), 617–634.

Lucas, M.; Lucas, M. The Difference Between Bitcoin and Blockchain for Business - Blockchain Pulse: IBM Blockchain Blog, 2019. Retrieved from https://www.ibm.com/blogs/blockchain/2017/05/the-difference-between-bitcoin-and-blockchain-for-business/

Nakamoto, S. Bitcoin: A Peer-To-Peer Electronic Cash System, 2008. http://bitcoin. org/bitcoin. pdf.

Skarmeta, A. F.; Hernandez-Ramos, J. L.; Moreno, M. V. A Decentralized Approach for Security and Privacy Challenges in the Internet of Things. In *2014 IEEE world forum on Internet of Things (WF-IoT)*, IEEE, 2014, pp. 67–72.

Ukil, A.; Bandyopadhyay, S.; Pal, A. IoT-Privacy: To Be Private or Not to Be Private. In *2014 IEEE Conference on Computer Communications Workshops (INFOCOM WKSHPS)*, IEEE, 2014, pp. 123–124.

What is the Bitcoin Blockchain? 2019. Retrieved from https://support.coinbase.com/customer/portal/articles/1819222-what-is-the-blockchain-

CHAPTER 4

Blockchain and Bitcoin Security

HOANG VIET LONG* and TONG ANH TUAN

People's Police University of Technology and Logistics, Bac Ninh, Vietnam

Corresponding author. E-mail: longhv08@gmail.com

ABSTRACT

Launched in 2008, blockchain technology with a typical product of Bitcoin is being applied and has great development potential in many areas such as finance, trade, manufacturing, and telecommunications. Bitcoin as a popular cryptocurrency with the highest total market capitalization of the virtual money market has always been an attractive target for hackers. In this chapter, we focus on attack and security on blockchain technology. In the first section, some targeted attacks on Bitcoins will be repeated. In the next section, we present potential attack techniques on blockchain technology in general and Bitcoin in particular, such as double-spending attack technique on mining pool or Bitcoin network. The general security solutions for blockchain will be presented in the next section. Besides, we devote a section to presenting the issue of how to secure wallets and Bitcoin transactions for those who use this virtual currency. The summary and discussion of future security orientation for blockchain technology will be presented at the end of the chapter.

4.1 REVIEW SOME ATTACKS ON BITCOIN VIRTUAL CURRENCY

When it comes to blockchain, many people will immediately think of Bitcoin, which is considered a typical and the most popular application of blockchain technology today. With Bitcoin, financial transactions can be authenticated and completed without the need for an intermediary, as well as not disclosing the identities of both parties. Bitcoin, with its advantages, sometimes reached 13,412.44 USD/Bitcoin in January 2018 (Wilmoth,

2019). This is an attractive bait for every hacker. Was there a successful invasion of Bitcoin and Bitcoin's blockchain technology that helped it fight traditional attacks? In fact, in the past few years, there have been many attacks on Bitcoin currency, especially when the currency has skyrocketed. In this section, we will consider that (Bhulania and Rai, 2018).

4.2   TWO ATTACKS ON THE TRADING FLOOR MT. GOX

In 2011, two years after the birth of Bitcoin, Mt. Gox, the largest trading platform of the time, based in Japan, was attacked to usurp Bitcoin, which did not happen just once, but twice. For the first time, on June 19, 2011, hackers somehow managed to get the Mt. Gox floor auditor's credentials, helping them transfer 2609 BTC to a Bitcoin wallet, for which Mt. Gox didn't have a key. This meant that they neither knew where they had transferred nor could get it back. More than 2600 BTC lost, causing the operation of Mt. Gox to be suspended for several days. They overcame the problem shortly later and continued to work on the market (Biswas and Muthukkumarasamy, 2016).

At that time, the amount of Bitcoin lost and its modest value may not be enough to create scenes with executives. Then they let the second attack happen in 2014, at the time Mt. Gox handled about 70% of Bitcoin transactions around the world. This time, the number of leaked BTC completely sank Mt. Gox's business. The players on this floor are dumbfounded, they don't know where their money is and when they can get it back.

Soon, Mt. Gox halted operations and filed for bankruptcy, claiming that more than 750,000 BTC (about $350 million at the time) were stolen and could not be recovered.

4.3   THE ATTACK ON BITFLOOR PLATFORM

A year after the first attack on Mt. Gox, another exchange attacked was BitFloor..

Date of attack: September 2012
Amount robbed: 24,000 BTC

On each system, private keys that send users to their wallets need to be encrypted to ensure privacy and avoid being exploited when they fall into the hands of hackers. The mistake of BitFloor comes from their administration, when the online users' private keys were not encrypted and somehow fell into the hands of the hackers. Hackers have used this data to steal 24,000 BTC from the user's wallet. Although BitFloor was able to make up for its users, the founder of the platform, Roman Shtylman, had to declare a temporary halt to the operation then (Bitcoin Historical Data | CoinMarketCap, 2019).

4.4 THE ATTACK ON BITFINEX PLATFORM

Another attack on Bitcoin was made at Bitfinex in Auguest 2016. Hackers paired nearly 120,000 BTC, worth nearly $72 million at the time.

Date of attack: August 2016
Amount robbed: 120,000 BTC

The attack was caused by a flaw in Bitfinex and BitGo's multi-signature architecture. By exploiting them, hackers accessed and retrieved 120,000 BTC. Bitfinex's multi-signature architecture was split with three pieces of private key, in which Bitfinex retained two pieces, the third was delivered to the security partner BitGo. Just have two or three pieces in hand to access and use Bitcoin. Hackers who hacked into Bitfinex system took a piece and then demanded BitGo to provide the remaining piece to take over the account (Bitcoin (BTC) Historical Data | CoinMarketCap, 2019).

Before the attack, Bitfinex used to be the largest exchange company for Bitcoin. Bitfinex believed that relying on the users to make up for the loss was the best way to avoid bankruptcy. This floor also hopes to reimburse the victims for damages. Bitfinex quickly solved the problem by issuing a token to the victims, making it possible to exchange in USD and, therefore, most of the Bitfinex investors have gradually been refunded under a certain schedule.

4.5 SOME OTHER ATTACKS

Besides attacks on major exchanges, many of the smaller-scale attacks have also frequently taken place since Bitcoin was born so far. The table below lists the time and losses of those attacks (CyptoCompare and various, 2017).

Time of attack	Target/trading platform	Amount of Bitcoin stolen	Estimated damage
October, 2011	Bitcoin7	5.000 BTC	$25,000
March, 2012	Bitcoinica	43.554 BTC	$215,000
December, 2012	BitMarket.eu	18.788 BTC	$236,000
May, 2013	Vircurex	1.454 BTC	$170,000
July, 2014	Cryptsy	11.325 BTC	$7,200,000
March, 2015	KipCoin	>3.000 BTC	> $777,000
May, 2016	Gatecoin	250 BTC	$113,000
April, 2017	Yapizon	3.816 BTC	$4,700,000

4.6 DISCUSSION

Ever since Bitcoin and other virtual currencies are based on blockchain, successive attacks have taken place leaving users unaware of Bitcoin's security and blockchain.

However, looking back at the attacks in the history of crypto coins, we will see that the flaws are not located in Bitcoin or other cryptocurrency, they are only high-value items and a target of craving want of hackers. In contrast, the risk of attack comes from weakness in the operation, management of exchanges, and e-wallet services. Indeed, every trading platform has its own weaknesses, and no perfect process can guarantee absolute safety, because no matter how much or less people still need human involvement. Saying so does not mean that Bitcoin or blockchain has no weaknesses. In the next section, we will consider which techniques can be used to attack and exploit this currency (Khatwani, 2015).

4.7 SOME ATTACK TECHNIQUES AT BITCOIN AND BLOCKCHAIN

Although there are still many debates about legitimacy, Bitcoin is increasingly popular and has a stable number of participants. Bitcoin development team as well as Bitcoin players always try to solve and minimize the risks that can happen to this virtual currency. So how can a hacker attack? The content of this section will present some techniques on how to attack Bitcoin system with potential threats to this currency trader (Goguen and Campbell-Verduyn, 2017).

4.8 DOUBLE-SPENDING ATTACK

Introduction about double spending

One of the most common types of fraud when using Bitcoin or other virtual currencies is double-spending. This is a form of fraud using two different transactions to spend the same balance in the account. That is, with the amount of X, if the attacker normally buys only one product that has a value of X, now they can buy two products of the same X value.

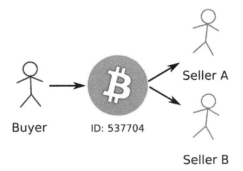

FIGURE 4.1   An illustration of an attacker using the same amount of BTC to buy items from sellers A and B.

To understand more clearly, let's consider a simple example as follows: I went to the auto shop and ordered a $300,000 Lamborghini supercar, making a cash payment. The transaction is complete, now my $300,000 is in the store's safe. Apparently I can't use that money to spend on another item right now (Hashrate Distribution, 2019).

Also in this case, but I opted to pay with Bitcoin instead of cash, and the Lamborghini mentioned above was priced at 100 BTC. I noticed another Rolls Royce next door that cost 100 BTC. Fortunately, Bitcoin is a digital currency, not physical cash that can be seen or directly manipulated on it. With my double-spending tricks, I can create a Bitcoin transaction to buy Lamborghini and copy and replay it to buy the next Rolls Royce battle. This helps my 100 BTC amount be spent up to twice. That is the scenario of double-spending attack: Two transactions coming from the same person with the same balance are transferred to two different destination addresses, resulting in two successful transactions on the same amount of money.

However, Bitcoin is a cryptocurrency with its own verification mechanism and the fact that it is quite difficult to carry out a double-spending attack on the Bitcoin system, to better understand this we learn in the next content of the chapter.

Double-spending attack

Double-spend attack is a technique that takes advantage of the 51% vulnerability in the blockchain network. We need to know that blockchain is a decentralized network. In it, a ledger to store data, usually Bitcoin is data about transactions. These data are contributed, validated, and managed by many yearly excavators scattered throughout the world. These peaches are the computer systems involved in decoding the blockchain network. A transaction is determined to be valid when more than 50% of the network members agree to authenticate it (Gervais et al., 2016).

Therefore, this network can operate and authenticate without a third-party intermediary to organize and manage. The data on the blockchain is public, transparent, and cannot be modified by any individual. However, the question is: If someone captures more than 50% of the network members, can they fully verify the transactions in the network at their own discretion or not?

How do I transfer BTC to buy a car?

In the above-mentioned example, when I made a transaction to transfer 100 BTC to the purse of the store to buy a Lamborghini, this transaction was encrypted and sent to a place called "unconfirmed transactions." Miners will choose these unconfirmed transactions and set up a block. Later, the block will be added to the blockchain network. However, in order to do this, miners need to solve a very complicated problem and that is why Bitcoin excavators require powerful computing capabilities. A lot of miners will compete with each other, who will solve their previous problem to get their block into the blockchain (Bitcoin (BTC) Historical Data | CoinMarketCap, 2019).

Not yet, the transaction in this block will be publicized and should be verified by other miners. If there are more than 50% validations, this block is shown on the blockchain and the transaction is recorded in the ledger. I officially completed the transfer of 100 BTC to buy my Lamborghini.

It should be noted that miners can only confirm transactions and block mergers into blockchain, but cannot create a transaction themselves, because they need to have a cryptographic signature of the Bitcoin storage account. Of course, only I can sign my 100 BTC storage account, so no one can access and steal my 100 Bitcoin (Khan and Salah, 2018).

I will double my 100 Bitcoin by making a 51% attack.

We will explore further to find out how a 51% attack works. As explained earlier, when a miner solves the problem to pair his block with the blockchain network, this block will be publicized for other miners to confirm and reach consensus with the required ratio above 50%.

However, a miner may not publicize his block on the network, and create a child blockchain that is not declared. We will split into two types of blockchain: The first is the green blockchain, which is authentic and is confirmed by the majority of miners; the second type is the orange blockchain—fake blockchain that is not declared.

How can I use my 100 BTC twice? First, I created a fake, unlisted, fake blockchain and a really blue blockchain.

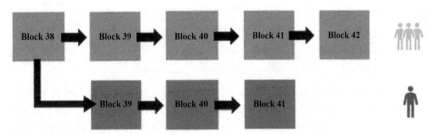

When I made a Lamborghini purchase with 100 BTC, I declared and confirmed the transaction on the real blue blockchain. This transaction is confirmed by other miners, and my wallet is deducted by 100 BTC.

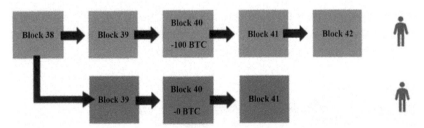

At the real blockchain, my transaction was verified and minus 100 BTC was found in my wallet. However, at the fake blockchain, I don't implement this transaction. That means, 100 Bitcoin is still in my wallet. This blockchain is not public, so other miners cannot know.

As shown in the image above, we can see the main difference is in block number 40. On blockchain color, the number 40 has been deducted with 100 BTC, but this does not happen in orange block 40, is the block public. Then I copied the next block of the blue blockchain and performed it myself on my orange blockchain. For example, in the figure is the block number 41, 42, and so on.

This is when trouble starts to happen.

Blockchain operates according to a majority of opinion-based governance model, which always confirms the longest blocks of blocks. Normally, the real blockchain version with the participation of many excavators around the world will have the fastest new block speed. This is how the blockchain determines which version is real (Vukolić, 2015).

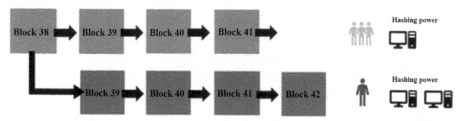

With some strong potential, I have the computing power equivalent to 51% of the entire excavator network. This can help me make my fake block-chain have faster validation speeds. The race started. If I have the power to handle faster than the real blockchain, I can turn my fake blockchain into a real blockchain. Thus, after creating longer blockchain chains, I broadcast my orange blockchain chain to all other miners. The important thing is I have to create a longer blockchain to gain the miners' priority.

The rest of the work will be for miners, they will discover that my fake blockchain is longer than the version they are working on. These miners are forced to move to work on the new blockchain and recognize its correct-ness—this is the chain I created (Wilmoth, 2016).

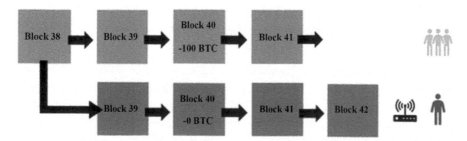

So, I fooled all the other miners, turning my fake blockchain into a real blockchain. And at block 40 of the fake blockchain chain, I still have my 100 BTC intact. I used it freely to buy the remaining Rolls Royce. Of course this time, I let the transaction proceed normally. And what the car shop received was only 100 BTC for two super cars. All like a clever magic show.

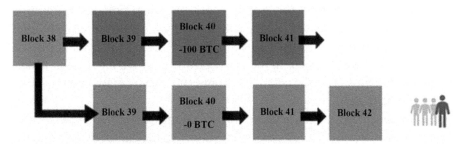

The above-mentioned scenario is called a 51% attack, when hackers try to exploit the 51% vulnerability on the blockchain network. In order to get a speed of validation faster than the entire blockchain network, I myself must have the power to calculate by 51% the power of all the excavators in the world to join this network, a first private is not small.

It sounds pretty easy to do a 51% attack, or double-spending. But that is only theory. There are not many 51% attacks that are clearly recognized by protection—shows the advantages of Bitcoin in particular and blockchain in general today.

4.9   ATTACK ON MINING POOL

What is mining pool?

Bitcoin is a peer hierarchy and no third party stands out as an intermediary or controller. Digging Bitcoin will help us confirm transactions from A to B, or prevent A from committing fraud when trading at the same time with B and C (double-spending). In Bitcoin's blockchain network, every miner has to compete with other miners to become the first to solve the problem the system offers. Miners will be rewarded with a bit of Bitcoin corresponding to the transaction fee they process. At the same time receive additional rewards for each Bitcoin block exploited, usually 12.5 BTC/block (Hashrate Distribution, 2019).

Because the reward for each Bitcoin block is too high, the competition among miners becomes more intense. There are hundreds of thousands of supercomputers looking to exploit the next Bitcoin block. Estimate the current blockchain's Bitcoin mining capacity with the world's top 500 super-computers combined and multiply by 1000 times. The constant competition of miners helps to increase the strength of the Bitcoin network, an important element to ensure the survival of the system (Lin and Liao, 2017).

Mining pool is a server that combines the power of excavators to create one or more algorithms to help users find their coins more easily. The strength of the excavator used is calculated by Hs (Hash Rate) (Bitcoin (BTC) Historical Data | CoinMarketCap, 2019). When synthesizing power from the pool, there will be a tremendous power to solve algorithms faster and more efficiently, then the rate of receiving BTC rewards for creating new blocks will be higher.

The mining pool, which can be called mining groups or excavations, is created to increase the computational power, which directly affects the verification time of a block, thus increasing the chances of gaining. Bitcoin rewards from the exploitation. When participating in the mining pool, the miners will contribute their processing power to a large system, thereby increasing the probability of receiving the participants' rewards and pool to share the profit for the members based on each person's effort.

For this purpose, in recent years, a large number of mining pools have been created and studied to exploit effectively. Figure 4.2 shows the market share until January 2019 of the most popular mining pool.

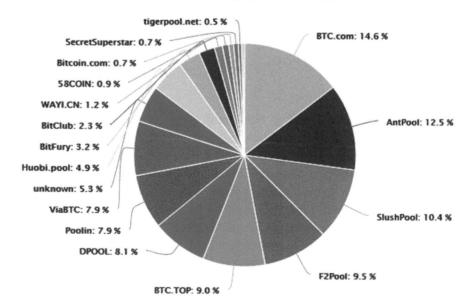

FIGURE 4.2   The market share chart of the most popular mining pool Bitcoin mining as of January 2019.

Source: Hashrate Distribution. Retrieved from https://www.blockchain.com/pools. 2019.

Most mining pools are managed by group managers, whose task is to forward unresolved "work" units to team member (Khan and Salah, 2018). The miners create partial proofs-of-work (PPoW) and Full proofs-of-work (FPoWs) and send them to the manager in the form of shares. When an operator discovers a new block, it will be sent to the manager with FPoW. The broadcast manager of that block in the Bitcoin network receives mining rewards, this reward is distributed to participating miners based on the percentage of shares contributed or in other words, participants are rewarded based on PPoW. The Bitcoin network now includes solo mining tools, open groups that allow any mining tool to participate.

Pool hopping attack

In recent years, the trend of exploiting exploits in the mining pool has been increasing. For example, dishonest miners can perform a variety of internal and external attacks against a mining group. Internal attacks are when operators in the group act maliciously to collect more of their job reward or break the function of the group to avoid successful exploitation efforts. In external attacks, miners can use their higher hash power to perform attacks like double-spending.

One of the attacks on Bitcoin is pool hopping. Pool hopping is like traveling in big cities where you can go anywhere on a tour bus and then sit on another bus, miners thank to a special software that can jump/move between the pool for the purpose of receiving greater remuneration for handling transactions on shorter blocks. Most miners use this software because it is difficult to monitor several pools themselves (Gervais et al., 2016).

Pool hopping makes continuous exploitation in a pool very unfavorable, miners only stay when the attraction is high and leave when gravity decreases. This damages the general attraction of participants of continuous exploitation.

Block withholding attack

Block withholding is another form of selfish mining. Suppose that you are a miner in the Bitcoin network using the ASICS chip with the largest computing power. Then, exploiting the blocks is an extremely easy task. However, suppose that when you dig a block, instead of announcing to the entire network and receiving the reward, keep this block and continue exploiting alone on the blockchain with this new block or keep the block with the item. For other purposes, this can be used for two types of attacks: sabotage and line-in-wait attacks.

To put it simply, block withholding is made when a group or an individual to retain a new block is found from the knowledge of others. The attacker then tries to exploit the secret and let the remaining operators work on one block to end the orphan blocks, causing them to waste resources without producing results. If the attacker can find the second block before the remaining operators find the first block, "he" will have an important advantage over the rest of the network (Bitcoin (BTC) Historical Data | CoinMarketCap, 2019).

As can be seen in the figure, by keeping the block(s) exploited, these operators deliberately branch the blockchain, which is considered selfish mining. They continue to exploit on their own chains, while other "honest" miners are still exploiting the public chain. If selfish exploiters can lead more on their own chains, their chances of earning more bonuses and the waste of honest exploiters' resources will also increase. At that time, honest operators will lose their reward or will not be paid according to the effort. Therefore, the influence of this selfish exploitation will be disastrous (CyptoCompare and various, 2017).

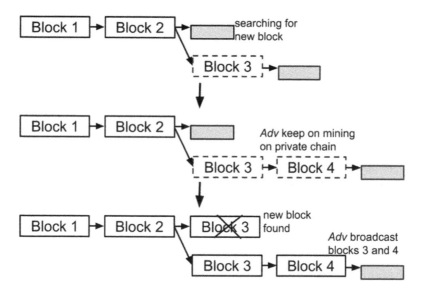

Sabotage

This is an attack technique that destroys a pool in which an attacker never sends any blocks. This has no direct benefit to the attacker, it only harms the pool operators and their participants.

Accordingly, each participant (including the attacker) will lose his reward with the corresponding reward for the computing power of the attacker in the group. If the attacker's hash rate is h, the total pool is H, the pool fee is f and the whole pool reward is pB then the participant will get an average $(1 - f)$ $(1 - h/H) pB$ per share. This can cause a significant loss to the pool, but since it is difficult to detect, it may not cause participants to leave the pool or cause any long-term disruption.

In some cases, this type of attack also puts the operator at great risk and leads directly to the disintegration of a pool. On average, the operator will receive $(f - h/H) pB$ per share sent and since f is usually only a few percent, the number that the operator receives is completely negative if the attacker's hash rate is higher (Dorri et al., 2017).

Lie-in-wait

The BTC reward on each block is cut in half after every 210,000 blocks and the transaction fee changes based on the existing transactions in the network. Most often, the mining pool allows participants to tap with them using a public Internet network, such groups are vulnerable to various security threats. Attackers believe that by not being "honest" about the number of shares contributed to each pool, the maximum reward will be earned.

Unlike sabotage, a lie-in-wait is a type of attack that causes an attacker to gain the maximum benefit from the pool's reward split, in which the attacker participates in various pool and divides the computing power. My access to those pool and use of knowledge about imminent blocks to focus on exploiting where possible bring the highest benefits (Wilmoth, 2019).

P+Epsilon attack

A proof-of-work system can be affected by a special attack called "attack P + Epsilon." To understand how this attack works, we must first define some terms.

- Non-coordinated selection model: Is a model where all participants have no motivation to work together. Participants can form a group but are never a large enough group to become a majority.
- Coordinated selection model: A model where all participants join together for the common good.

Bribery attack model

With an uncoordinated model, what happens when an attacker enters the system and gives incentives to bribe the miners to work together? This new model is called bribery attack model. To succeed in this model, the attacker needs to have two resources: (Goguen and Campbell-Verduyn, 2017).

- Capital: The total amount of money that an attacker owns to be willing to pay for making miners perform a specific action.
- Cost: The price that the miner actually received after taking action.

However, if the attacker decides to attack on the blockchain, we will face an interesting problem called "P + Epsilon" attack. Consider the Table 4.1 below:

TABLE 4.1

		You vote 0	You vote 1
Base game	Others vote 0	P	0
	Others vote 1	0	P

Imagine a simple scenario in an election. If people vote for a specific person in the same way that others vote, they will receive a reward and otherwise receive nothing. Now if an attacker enters the system and gives the opposite condition to an individual. If you vote AND others do not vote, you will receive the "P + ε" reward. Regular bonuses and a bribe add ε on it.

So now, the payment matrix looks like Table 4.2:

TABLE 4.2

		You vote 0	You vote 1
With bribe	Others vote 0	P	$P + \varepsilon$
	Others vote 1	0	P

Now imagine everyone involved in this scenario knows that if they vote, it is possible that they can receive the bonus, but if they do not vote, there are 50–50 chances of getting paid money.

What do you think participants will do in this case? Of course they will vote to get a guaranteed refund. Now this is where things get interesting. As can be seen in the matrix, the registrant only has to pay the bribe "ε" when only one person votes while others do not. However, in this situation because everyone voted, balance turned (Table 4.3):

TABLE 4.3

		You vote 0	You vote 1
With bribe	Others vote 0	P	$P + \varepsilon$
	Others vote 1	0	P

The attacker doesn't even need to pay bribes! Therefore, approach this issue from the perspective of the bribe: (1) Convince the group to vote in a specific way; (2) Achieve goals without actually paying bribes (Johnson et al., 2014).

This is a big win–win scenario for the bribe and this has a heavy impact on the blockchain system, especially in a work-evidence system. Check out our old hypothetical blockchain again:

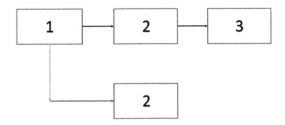

Suppose the bribe really wants a hard-branched chain and declares that a group of miners who switch to a new chain will receive a bribe of ε, which will be the driving force for the entire mining team coordinated and joined the new chain.

Clearly, bribery costs must be extremely high so that such a thing can happen, however, as we have seen in the above bribery model, the attacker will not even need to pay the above-mentioned amount. According to Vitalik Buterin, this is one of the biggest problems of the proof of work, which is its vulnerability to P + Epsilon attack.

Blacklisting

Blacklisting is a form of attack targeting a single entity. Imagine that there is a prominent entity that has a lot of capital wants to perform an attack on an object named Jake by having this guy blacklisted online. In fact, suppose that this attack entity is a country that controls a very large exploitation capacity, here we call the country Aliceland. Symbols of blocks are as follows:

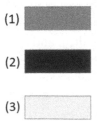

(1) Common block
(2) The block dug by Aliceland
(3) Jake's trading blocks or Jake's block

There are two ways that Aliceland can give Jake to the blacklist:
- Punitive forking
- Feather forking

Case #1: Punitive forking

There are two strategies that can be used to perform punitive forking.

Strategy 1:

Since Aliceland was a country, Aliceland simply did not put any of Jake's transactions in the block. In fact, this is a naive strategy because to do it needs to consider many things. They will need to control 100% of excavations in the country. As long as a mine is out of control, Jake's transactions can be leaked and put into the block.

Strategy 2:

Aliceland wanted to do punitive forking and own over 51% of the world's mining power. They will notify the world that they will not allow any transaction containing Jake's transactions in the blockchain. In other words, the moment a Jake's block appears in the blockchain, Aliceland will start creating a branching.

(1) Appeared a Jake's trading block in Blockchain
(2) Aliceland started a branching operation without agreeing with other members because it owned 51% of computing capacity.

Since they announced their previous intention and owned 51% of the majority, no one dared to oppose them and even when fighting and accepting a block containing Jake's transaction, they would be forced to create blocks, orphan block because Aliceland then simply needed to use the majority of the power to separate and continue on the new branch without Jake's block. (CyptoCompare and various, 2017).

Case #2: Feather forking

In Case #1 we have seen the way Aliceland performed to remove some object with punitive forking when accounting for 51% of the network's computing power. The question now is how can Aliceland add Jake to the blacklist without owning 51% of the majority? In this case, Aliceland can put Jake on the blacklist using a method called feather forking.

Aliceland can declare that whenever a block of Jake's transactions is added, they will try to branch, but will give up after a certain number of assertions. This is different from punitive forking, Aliceland does not threaten permanent branching, and instead, they say they will branch but will also return after a certain number of assertions.

4.10 BITCOIN NETWORK ATTACKS

Distributed denial-of-service

We will start with the current popular network attack technique, distributed denial-of-service (DDoS) attack. The targets of these attacks are Bitcoin currency exchanges, mining pools, eWallets or other financial services in Bitcoin (Dorri et al., 2017).

DoS and DDoS are techniques that an attacker tries to make a certain server or device inaccessible by flooding the target with more requests than the server can handle. Bitcoin network with distributed property and consensus protocols, launching a normal DoS attack is less feasible and does not affect the network functions much. Hackers try to make as powerful DDoS attack as possible to aiming tamper with network tasks. Unlike DoS, in which there is only one person performing, DDoS attack controls a Botnet network and launches multiple sources of attack at the same time to the target. Although it does not affect the data, this attack causes congestion and stagnation of the network, which can cause significant errors and damages (Vukolić, 2013).

In the Bitcoin network, successful DDoS attacks can help attackers eliminate "competitors" in conducting "mining" Bitcoin.

When the attack takes places, network resources are exhausted, preventing legitimate user access to the system. For example, an honest operator is congested with requests (like fake transactions) from a large number of customers created by the attacker's control. After a while, honest operators will be able to begin eliminating all incoming requests including requests from legitimate users. According to a comprehensive empirical analysis of DDoS attacks on Bitcoin networks by recording key events: 142 unique DDoS attacks on 40 Bitcoin services and 7% of all known operator's victims of these attacks (Bitcoin (BTC) Historical Data | CoinMarketCap, 2019).

Bitcoin transaction malleability

Transaction malleability is an issue of Bitcoin for many years. It is popular and has often become the target of hackers. In fact, there were many blocks in blockchain that were affected by this problem. They are considered as a major problem that requires remedial action.

We look back at how a Bitcoin Transaction works before going into the details of the issue: Users initiate an application, then sign and send a payment request to the Bitcoin network, and wait until the transaction is "dug" by the miners. The transaction is automatically assigned to a transaction ID for identification (Wilmoth, 2014).

In malleability, a transaction can be changed a bit and it can still be noted, everything still seems to go exactly like what the originator expected and does not change the signatures. This small change will cause the transaction ID to change because the transaction ID is created based on the content of the transaction.

Transaction Malleability

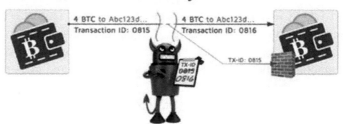

The problem is that when an organization/individual executes a transaction, it will be impossible to ensure that their transaction retains the transaction-ID value within the authentication waiting period or not. This is also the reason that Mt. Gox announced that their users lost Bitcoin when the exchanges performed the transaction monitoring and management based solely on transaction ID.

In the case of Mt. Gox, they were attacked by transaction malleability technique. Therefore, they are forced to suspend, withdraw and freeze customers' accounts. The attack on which Mt. Gox is a victim proceeds as follows:

(i)   A dishonest customer C_d sends n coins to his Mt. Gox account.
(ii) C_d sends a transaction T to Mt. Gox asking to send back to C_d n coins.
(iii) Mt. Gox made an T_0 transaction to transfer n coins to C_d.

(iv) C_d implements a malleability attack, receiving the T_0 semantically equivalent to T but different transaction ID, now assuming that T_0 is included in T's replacement blockchain.

(v) C_d complained to Mt. Gox that the T transaction failed.

(vi) Mt. Gox performs an internal audit and will not find a successful transaction with T's original transaction ID, so Mt. Gox refunds the user's wallet. Since then, C_d can make withdrawals twice.

The whole problem is in Step (vi), where Mt. Gox should search for transactions with the original transaction ID of T and not T_0.

4.11   TIME JACKING ATTACK

Another attack technique on Bitcoin is time jacking attack. In Bitcoin network, all participating nodes maintain the network timer. The timer's value is based on the average time of the button's colleagues and it is sent in the version message when first connected. However, if the average time is more than 70 minutes from the system time, the network timer will return to the system time.

An attacker can create many fake colleagues in the network and all of these colleagues will report incorrect timestamps. This can slow down or speed up the network timer of a node. An advanced form of this technique involves speeding up the clock of most miners while slowing down the target's clock. Since the time value can be deviated by up to 70 minutes, the difference between the times of the nodes will be 140 minutes. Furthermore, by notifying the timestamp correctly, an attacker can change the network's timer and deceive it to accept an alternative blockchain (Khatwani, 2010).

This attack technique can lead to the following risks: The success of double-spending attack, depleting miners' computing resources, and slowing the speed of transaction confirmation.

4.12   OTHER THREATS TO BITCOIN AND BLOCKCHAIN

In the blockchain products, Bitcoin is still the most noticed name. In addition to the above-mentioned issues, Bitcoin users in particular and blockchain in general also face other potential risks.

The increase in the popularity of Bitcoin has attracted more new users. Each person owns a set of public and private keys to access their account or wallet. Revealing these key pairs may cause the account to be misappropriated and misused. Therefore, administrators need to have the management techniques to be safe. In order to use Bitcoins, users need to install wallets on their desktop or mobile device. Wallet theft is mainly done by mechanisms including hacking the system, installing software bugs, and using inaccurate wallets.

Bitcoin uses the Elliptic Curve Digital Signature Algorithm (ECDSA), which is standardized by NIST to sign transactions. For example, consider the Pay-to-PubKeyHash Standard trading script (P2PKH) that users need to provide a public key and signature (using their own key) to prove wallet ownership. To create a signature, the user chooses a random value for each signature. This value must be kept confidential and it must be different from all other transactions. Repeat the value on each signature to increase the likelihood of calculating a private key. Therefore, it is necessary to increase the security of ECDSA to use highly unique and random signature values for each transaction signature (Bhulania and Rai, 2018).

It can be seen that the key pair for security for Bitcoin wallet is often hackers' target. This raises the issue of storage and security key management. Recently, different types of wallets have been developed and studied to have a mechanism for storing safer keys for users.

4.13   SECURITY ISSUES FOR BLOCKCHAIN AND BITCOIN

Over 10 years of blockchain and Bitcoin development, developers have also gradually overcome the weaknesses of this technology. In this section, we will discuss some of the security issues for Bitcoin and blockchain today.

Why is a 51% attack only in theory?

Previously, pre-coding experts believed that a 51% attack was only in theory. In fact, achieving power equivalent to 51% of all excavators in a blockchain network is impossible.

For Bitcoin, experts believe that this currency can never be affected by a 51% attack. Even the most powerful computers in the world today cannot compete directly with the total computing power of Bitcoin's blockchain network.

Suppose you want to do it, first you will need huge factories and the most powerful supercomputers. Of course, your behavior will not be able to blind the managers, it is easy to discover someone that is capable of acquiring 51% of the power of a large blockchain network like Bitcoin.

In other blockchain systems with smaller scale, the same thing is more likely to happen. For example, the 51% attack on Verge in April 2018 (Verge,

2018). The main reason for the exploit is due to a vulnerability in this block-chain network, allowing hackers to create new blocks with lightning speed without the need for actual computing power to be 51% of the entire network (Lin and Liao, 2018).

In addition, experts said that a blockchain with a small scale of a certain altcoin could be attacked 51%. This may be because smaller networks will need less computing power than attacking Bitcoin networks. That's why we often see other cryptocurrencies that are often attacked by 51%, which has never happened to Bitcoin.

It can be seen that 51% attack is an inherent weakness in every blockchain system. The threat of this vulnerability decreases as the size and computing power of the blockchain network increases. In fact, the use of this technique to attack valuable blockchain networks always requires huge potentials that are hard for any attacker to meet.

4.14   ASIC EXCAVATORS CAN SUPPORT 51% ATTACKS

A topic that is increasingly being concerned by the cryptocurrency commu-nity is the explosion of ASIC excavators. These are specialized excavators, with much higher computing power than regular graphics card computers. These ASIC excavators make power focused on the hands of some indi-viduals who own them (Bitcoin (BTC) Historical Data | CoinMarketCap, 2010).

A typical example is Monero's blockchain (XMR), which recently imple-mented an update that prevents the ASIC excavators from taking part. The results were unexpected when the total power was reduced to 80% (Goguen and Campbell-Verduyn, 2018). Of course, this 80% doesn't just focus on an individual, but this result still worries many people. Bitmain—the world's largest ASIC maker—is suspected of controlling a large number of crypto-currency operations with these ASIC excavators.

The Bitcoin Gold development team has also recently suffered a 51% attack, also announced it will release an update to remove ASIC excavators from participating in its network.

Currently, it can be concluded that no implemented solution can guar-antee the complete removal of the 51% risk in the blockchain. This is still a possible threat and can happen at any time, when the gain is not greater than the cost.

4.15   HOW DOES PROOF-OF-WORK HELP PREVENT DOUBLE-SPENDING?

Consider a case of the double-spending gap: Suppose Alice has an online store that accepts Bitcoin payments. Bob accesses this website and orders an iPhone. If Bob chooses Bitcoin as the payment method, Alice will of course wait for payment confirmation before sending the iPhone to Bob. But because of the way the block works, Bob can try to deceive Alice by initializing two signature messages with the same signature—one in which he sends money to Alice, and the other he sends to himself with another address.

When Alice sees the transaction message that Bob sent her money, she will deliver the product. However, if the transaction that Bob deposited for himself is put into the block before the deposit for Alice, then the transaction will be executed—that means Bob will get his money back and still have get the iPhone (Biswas and Muthukkumarasamy, 2012).

To avoid this, Alice may not send the iPhone immediately after the transaction is announced, instead, she will wait until it is on the blockchain, which means, Bitcoin has reached her purse. But that still is not a good choice. Because in some cases, there may be more than one block added at the same time, creating a fork in the chain. In these cases, the digger at the end of a block can choose which branch he wants to add. Pretty quickly, a branch will become longer than the other branch, the shorter branch will be removed and all transactions on it will be transferred back to mempool.

From this, we see how proof-of-work—performing more computing tasks—helps keep transactions safe (Gervais et al., 2016).

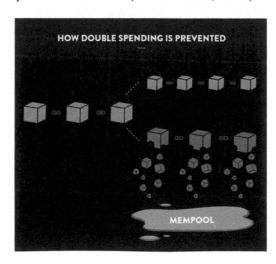

But let's consider another hypothesis: What if Bob could create two branches, with one block containing legitimate transactions and the other having fraudulent transactions? And he added that fraud transaction to the same rate that other diggers added to the other legitimate transaction. Alice sees the branch with legal transactions growing, will get a sense of security, make her send the iPhone. This may be wrong because Bob can make his string longer, sending legitimate transactions back to mempool. Because it is the same signature as a fraudulent transaction, so if it is reselected, it will be considered invalid.

That's an interesting hypothesis, theoretically it could happen. But in fact, that is quite poorly feasible. This is because it requires a large computing power and a small time to solve and add to a block. Even with extremely powerful processors, to do this, Bob will need to control more than half of the blockchain's computing power (Vukolić, 2012).

So, the proof-of-work makes fraud on Bitcoin online is much more impossible. Besides, mobilizing a huge processing capacity to conduct fraud seems to be ineffective. With that processing power, an attacker can legally mine Bitcoin and gain more BTC than intentionally stealing it.

4.16   WHAT DO BITCOIN AND ELECTRONIC MONEY USERS NEED TO DO TO PROTECT THEMSELVES?

Electronic money appearance brings a great potential, along with many risks and risks for participants. In order to protect themselves, users need to have the minimum security knowledge on the Internet environment. In this section, we will discuss how to reduce the risk of users taking part in trading electronic currencies in general and Bitcoin in particular.

Protect your private keys and money

On the Bitcoin network, owning this virtual currency means you have a corresponding wallet and private key. A key is a 256 bit data length (Goguen, M.; Campbell-Verduyn, 2014), which can also be represented by letters and numbers. An example of a private key is as follows:

"5KJvsngHeMpm884wtkJNzQGaCErckhHJBGFsvd3VyK5qMZXj3hS".

This private key allows users to encrypt digital signatures, access their wallets, and use the money in it. Without them, users have no way to prove that the number of Bitcoin in their wallet belongs to them. Losing private keys is like losing your Bitcoin. Therefore, the private key needs to be stored in the safest place (Gervais et al., 2011).

Password optimization

Password is one of the most sensitive information for each account. There are many people who set up simple passwords, such as 12345678, iloveyou. And with passwords of this type, hackers only need about 10 seconds to find it. Some suggestions to increase the complexity of passwords are as follows:

- Set password with length of 10 characters or more.
- Set password with capital letters, numbers, and special characters.
- Avoid setting a password that relates to your date of birth, interests, or personal information.
- Do not set a password for different accounts.
- Do not write passwords to paper or store in unsafe places.

Use software, operating system copyright

When investing in the cryptocurrency market, you must be very careful with your computer, you should not access strange websites, ads, images, or emails with unusual signs. Because sometimes antivirus software cannot protect you from troubles.

It is recommended to install one of the best antivirus software so you can be more secure when doing transactions on your computer. Do not download or use pirated software of unknown origin. This is the way hackers use to eavesdrop and steal information quickly.

Limit public network using

Using public network is also quite a serious problem that beginners of cryptocurrency often encounter. It could be logging in at cafes, Internet cafes, or at work establishments (Goguen and Campbell-Verduyn, 2012).

When using public networks, security features are often overlooked. Hackers can easily hack into the network and take away important information on the modem. Just capture packets or use some special tools, they can detect information related to the user's computer, including passwords or login information.

In addition, users can enter the public network trap created by hackers. Fishing techniques and social industries will be used to steal accounts. Therefore, using your home network and using a cable to connect to the network is a safer way to access your wallet.

Use two-factor authentication

Two-factor authenticator is the way to protect Bitcoin wallet, which is familiar and effective. In particular, to access the account, users will need two information to authenticate, which may include one of the following factors: Password, secret question, OTP code or fingerprint. All of these factors make forging more difficult.

Consider storing money offline

A security option is to store money and private keys on a USB drive. This ensures that hackers can hardly steal your information. However, if the user loses that drive, this is similar to losing your account information.

Another option is to trust a third party, a Bitcoin wallet provider. This is a type of software that stores addresses and key pairs for all Bitcoin transactions. However, the growing number of current attacks targeting cryptographic exchanges has made it unsafe and always threatened. The recommended way is to store offline keys (Hashrate Distribution, 2014).

4.17   DISCUSSION OF THE DEVELOPMENT DIRECTION FOR SECURITY IN BLOCKCHAIN

Since the blockchain and Bitcoin were launched in 2008, they have attracted the attention of the community, which is considered a technology with potential for application in many areas in the future. Parallel to that is the issue of security and technology perfection that are always concerned. Let's discuss the possibility of setting out with blockchain in the future.

Security model based on asymmetric encryption

Blockchain security model uses asymmetric encryption to identify, authenticate users, and authorize transactions. Each account will have a pair of public and secret keys. While the account number in the blockchain is used as a public key, the private key will be owned by the user. This helps them authenticate and access the account. Transactions with electronic signatures created with the corresponding private key are considered valid and users can perform transactions such as purchases and transfers.

It can be seen that the private key is the only security tool that allows users to authenticate their account. When the account's private key is given to another person or stolen, it is difficult to find clues to retrieve them. There are no additional security measures to protect assets on the blockchain system. Anonymity in blockchain transactions has many advantages, but that also makes users lack the ability to authenticate themselves when being stolen. It is absolutely necessary to keep the private key secret to them to protect their account (Wilmoth, 2014).

Asymmetric cryptography used in blockchain is considered one of the best and safest encryption methods available today. Combined with other features, security on the blockchain is ensured. However, not having an additional security plan to protect blockchain users from losing or stealing private keys is clearly a shortcoming of the blockchain. This makes the account lack a second layer of protection.

Technology and scalability are limited

Blockchain's peer-to-peer network helps solve two goals: On the one hand, it allows people to add new transaction data to the maintained history; on the other hand, it ensures that the history of transaction data is protected from manipulation or tampering (Hashrate Distribution, 2019).

As the blockchain grows, the amount of data is getting bigger and bigger, the exploitation of virtual money, making transactions is also getting more and more difficult, it takes more time to synchronize data. At the same time, data is continuously increasing, causing major problems for customers while running the system. Blockchain uses complex hashing puzzles and requires network members to resolve it. This security measure increasingly requires a lot of processing resources and is accompanied by limited scalability. This characteristic of blockchain is considered a serious obstacle when deploying its application in contexts that requires high processing speed, high scalability, and high throughput.

Threats to electronic money still exist

Cryptocurrency like Bitcoin especially emerged in 2017, then gradually tended to cool down. They face many concerns about legality and value instability due to speculation and security issues. The predictable major trends, which are the gradual decline of the number of virtual money miners, attacks on Bitcoin are increasing and the currency is used for illegal transactions.

Since 2013, Bitcoin mining has been extremely developed. Part of this enthusiasm is fueled by the expectation of an increase in the price of Bitcoin as well as other cryptocurrency (altcoin). However, with a peak value of nearly 20,000 USD/BTC in December 2017, Bitcoin dropped suddenly and is only stable at over 4000 USD/BTC (Wilmoth, 2019). This in addition to the cost, complexity of increasing calculations, the potential for the currency to have no hope makes the interest of the Bitcoin miners fall.

In the Blockchain world, Bitcoin is still an attractive target for hackers. At the moment, double-spending offensive techniques are difficult to implement. But when the miners gradually leave the field, the imbalance in exploitation power will be clearer, creating clearer conditions for this attack technique to take place. Attacks on trading floors and user wallets also make Bitcoin users worried. Without improvements in security, blockchain technology in general and Bitcoin cryptocurrency in particular, there will still be concerns from users (Bhulania and Rai, 2018).

The reason why Bitcoin can develop in the future is that it is very useful for illegal transactions. Ransomware attacks require victims to pay with Bitcoin, criminals who use Bitcoin as money laundering tools. That just brings development opportunities for Bitcoin, but also gives it the legal and underestimation of users.

4.18 SUMMARY

Blockchain with its potential has been applied to life, the most prominent product is Bitcoin cryptocurrency. Bitcoin is often targeted by hackers because of its high value, reviewing some of the attacks that have occurred: Attack on the Mt. Gox trading floor in 2011 and 2014, BitFloor floor attack in 2012, and Bitfinex floor in 2016.

To attack the blockchain, hackers can use techniques that include:

- Double-spending attack;
- Attack on mining pool; and
- Attack on Bitcoin network.

Because of the ever-expanding network of blockchain, 51% attacks are often unlikely. However, ASIC excavators can make this more feasible. Work evidence can help prevent these vulnerabilities, besides the necessary security skills of users of blockchain or Bitcon.

Discuss some of the security issues of blockchain and Bitcoin in the future:

- Blockchain security model based on asymmetric encryption;
- Technology and scalability are limited; and
- Security issues and electronic money futures.

ACKNOWLEDGMENT

This work was supported by the Domestic Master/ PhD Scholarship Programme of the Vingroup Innovation Foundation.

KEYWORDS

- **blockchain technology**
- **bitcoin**
- **cryptocurrency**
- **attacks on blockchain**
- **blockchain security**

REFERENCES

Bhulania, P.; Rai, G. Analysis of Cryptographic Hash in Blockchain for Bitcoin Mining Process. In *2018 International Conference on Advances in Computing and Communication Engineering (ICACCE)*; IEEE: New Jersey, 2018; pp. 105–110.

Biswas, K.; Muthukkumarasamy, V. Securing Smart Cities using Blockchain Technology. In *2016 IEEE 18th International Conference on High Performance Computing and Communications; IEEE 14th International Conference on Smart City; IEEE 2nd International Conference on Data Science and Systems (HPCC/SmartCity/DSS)*; IEEE: 2016, pp. 1392–1393.

Bitcoin (BTC) Historical Data | CoinMarketCap. (2019). Retrieved from https://coinmarketcap.com/currencies/bitcoin/historical-data/.

CyptoCompare and various, Reuter, Profr. Tyler Moore Univ. Tulsa, 2017. Link: https://www.cryptocompare.com/.

Dorri, A.; Kanhere, S. S.; Jurdak, R.; Gauravaram, P. Blockchain for IoT Security and Privacy: The Case Study of a Smart Home. In *2017 IEEE International Conference on Pervasive Computing and Communications Workshops (PerCom Workshops)*; IEEE, 2017; pp. 618–623.

Dorri, A.; Steger, M.; Kanhere, S. S.; and Jurdak, R. Blockchain: A Distributed Solution to Automotive Security and Privacy. *IEEE Commun. Mag.* **2017,** *55* (12), 119–125.

Gervais, A.; Karame, G. O.; Wüst, K.; Glykantzis, V.; Ritzdorf, H.; Capkun, S. In *On the Security and Performance of Proof of Work Blockchains*. Proceedings of the 2016 ACM SIGSAC Conference on Computer and Communications Security; ACM: New York, 2016; pp. 3–16.

Goguen, M.; Campbell-Verduyn, M. The Mutual Constitution of Technology and Global Governance: Bitcoin, Blockchains, and the International Anti-Money-Laundering Regime. In *Bitcoin and Beyond*; Routledge, 2017; pp. 69–87.

Hashrate Distribution. Retrieved from https://www.blockchain.com/pools. 2019.

Johnson, B.; Laszka, A.; Grossklags, J.; Vasek, M.; Moore, T. Game-Theoretic Analysis of DDoS Attacks against Bitcoin Mining Pools. In *International Conference on Financial Cryptography and Data Security*; Springer: Berlin, Heidelberg, 2014; pp. 72–86.

Khan, M. A.; Salah, K.   IoT Security: Review, Blockchain Solutions, and Open Challenges. *Future Gener. Comp. Sy.* **2018,** *82,* 395–411.

Khatwani, S. (2019). Top 6 Biggest Bitcoin Hacks Ever. Retrieved from https://coinsutra. com/biggest-bitcoin-hacks.

Lin, I. C.; Liao, T. C. A Survey of Blockchain Security Issues and Challenges. *IJ Network Security* **2017,** *19* (5), 653–659.

Tudor, N. Bitcoin Network is 100,000 Times More Powerful than the top 500 Computers. blockchainflashnews, no. Bitcoin, Tutorials. p. https://blockchainflashnews.com/ bitcoin-network-is.

Verge Victim to Yet Another 51% Attack, XVG Down 15% In Past 24 Hours. (2019). Retrieved from https://cryptoslate.com/verge-victim-to-yet-another.

Vukolić, M. The Quest for Scalable Blockchain Fabric: Proof-Of-Work vs. BFT Replication. In *International Workshop on Open Problems in Network Security*; Springer: Cham, 2015; pp. 112–125.

Wilmoth, J. Monero Hard Forks to Maintain ASIC Resistance, But 'Classic' Hopes to Spoil the Party. Retrieved from https://www.ccn.com/monero-hard-forks-to-maintain-asic-resistance-but-classic-hopes-to-spoil-the-party. 2019.

CHAPTER 5

Blockchain Internet of Things

SIDDHARTH SAGAR NIJHAWAN,[1*] ADITI KUMAR,[2] and
SHUBHAM BHARDWAJ[1]

[1]*Netaji Subhas University of Technology, New Delhi, India*

[2]*Sri Venkateswara College, New Delhi, India*

Corresponding author. E-mail: siddharth16871687@gmail.com

ABSTRACT

The interconnected network of smart devices with the aim of collecting data and making intelligent decisions is known as Internet of Things (IoT). With the advancement in technologies, this vision is turning into a reality. Although, we have taken a huge leap in IoT domain, there are certain challenges to be addressed particularly in the domain of security and reliability of data. IoT is vulnerable to threats concerning data security and privacy. To tackle this issue, blockchain has surfaced as a key technology, which has the capability to transform the way we share information. Blockchain follows "security by design" procedure that can help in addressing various security requirements of IoT. Blockchain properties, namely, auditability, transparency, immutability, data encryption, and operational resilience can be used to overcome various architectural shortcomings of IoT.

This chapter exhaustively investigates the integration of blockchain and IoT with the aim to study the current research trends in the usage of blockchain-related modules aiding IoT. A complete review on the ways blockchain can be adapted to the specific needs of IoT so as to develop blockchain-based IoT (BIoT) applications are presented. It also analyzes various challenges in integrating the two technologies and depicts analysis as to how blockchain could potentially improve the domain of IoT. At the end, the main open issues and future research directions are discussed.

5.1   INTRODUCTION

The services based on Internet of Things have seen a substantial growth in the world, especially in manufacturing industries and healthcare. Transformation of urban areas into smart cities is also a major application of IoT. According to Lund et al. (2014), over 30 billion devices are expected to be connected with the web of IoT by the next decade. It is expected to vastly contribute to the world economy along with improvising on the quality of life. With the help of this technology, a network of cheap sensors and interconnected devices can be deployed with ease to collect information with a much higher granularity. The mass collection of data improves the efficiency and hence delivers advanced services in a variety of domains. It is clearly depicted in Das and Maniklal (2015) that this enormous collection of data is rapidly increasing, which has caused various security as well as privacy issues. This data can be computed to construct a virtual biography of our day-to-day lives revealing our lifestyle and behavioral patterns. The risks are known to various manufacturers and researchers but such concerns remain neglected as described in Wurm et al. (2016). There is a lack of fundamental security measures in various IoT products present in the market, which exacerbates such privacy risks. Smart devices such as vehicles (Amoozadeh et al., 2015) and smart locks (Ho et al., 2016) have been the prime target of such vulnerabilities. Certain IoT features like presence of heterogeneous device resources and lack of centralized control amplify the security challenges faced by manufacturers. To make the situation even worse, IoT solutions are being deployed in many sectors: homes, accessories in the form of wearable devices, automobile industry, electrical grid systems, and so forth. Such systems generate large volumes of data and require continuous input of power and connectivity. This gives rise to further limitations in the form of computational power, memory, limited power supply, and networking. Besides, the resource management and heterogeneity challenges present in IoT, the trustworthiness of data being accumulated and transferred is an important issue as well (Ahlmeyer and Chircu, 2016).

IoT plays a vital role in digitization of the information but faces a big challenge of security and privacy along with reliability of the data. To tackle such problems, protocol known as blockchain can be applied to assure trustworthiness of the data due to its cryptographic security benefits. As the public record of data is secured by a network of peer-to-peer participants, it has gained popularity in various applications such as smart contracts, digital assets, and distributed cloud storage. Due to the increasing demand, manufacturers throughout the world are focusing on blockchain to take care of security and privacy issues of IoT. There are large number of research publications on blockchain-based IoT technology that deals with general applications of the blockchain in the field of IoT. Huh et al. (2017) demonstrated that configuring and managing IoT devices with the help of blockchain smart contracts is far more feasible than traditional methods. This eliminates the synchronization as well as security issues involved in the traditional client–server model. Research work conducted by Conoscenti et al. (2016) demonstrate the use of blockchain beyond cryptocurrencies and further facilitates their applications in IoT. Various government entities have been using the concept of blockchain in voting systems as depicted in Ahlmeyer and Chircu (2016) and Levitt (2017).

The key features of blockchain that contribute in making it an efficient technology for addressing various challenges in IoT are summarized as follows:

- Pseudonymous identities: This plays a key role in the field of IoT where the identity of users is kept private.
- Decentralization: Since the resources of all the participating nodes are used, there is a lack of central control thus improvising the robustness and scalability of the entire system and eliminating many-to-one traffic flows at the same time. This tackles the problem of single point failure.
- Enhanced security: Blockchain is realized using a secure network over untrusted parties, which is highly recommended in IoT consisting of numerous heterogeneous devices.

The integration of blockchain with IoT gives rise to various challenges as well; resource management is one of them. In majority of IoT devices, resources such as computational power, supply, and so forth are limited, whereas blockchain mining is a highly computationally intensive process. Blockchain architecture is time-consuming, which is a disadvantage as IoT applications require low latency. Since the bandwidth is limited in an IoT device, overhead traffic caused by blockchain protocols might be undesirable. The large number of nodes existing in an IoT network will cause blockchain to scale poorly (Bubley, 2017).

Keeping the challenges in mind, this chapter provides the technical details of blockchain and IoT with emphasis on efficient methods of integration of the two technologies. First, we discuss the technical aspects of a typical IoT infrastructure by discussing various layers present in the network. Then, we list the challenges faced by typical IoT architecture. Further, we provide an analysis of progress in blockchain technology detailing its advantageous aspects. The chapter elaborates the impact of blockchain protocol on IoT in domains of security, privacy, and resource requirements. The chapter also deals with some of the unique practical challenges to the integration of blockchain and IoT. Few of the existing blockchain IoT platforms are also listed and critical IoT-related blockchain issues are discussed in detail.

5.2   INTERNET OF THINGS

A typical IoT environment consists of different devices that are recognizable and addressable via a set of sensors. These devices can also be controlled over internet using communication networks that can be either wireless or wired. The key role of an IoT ecosystem is to permit various devices to be connected to anyplace, anytime, and accessible by anyone using any network or service (Ducklin, 2016).This section gives technical description of IoT including its architecture. It also describes how IoT is different from traditional networks. Various threats to the IoT system are also discussed.

5.3   IOT ARCHITECTURE

There are a number of architectures describing a typical IoT system and its functionality. The most popular architecture used for IoT is given by IoT World Forum (IWF) committee (Stallings, 2015). They devised a reference model that gives a common framework for deployment of IoT in the industry. It can be compared to the Open System Interconnection (OSI) model of networking as it consists of seven layers as described in Figure 5.1. IoT is converted from a conceptual model into a real feasible system using this architecture (Shah and Yaqoob, 2016).

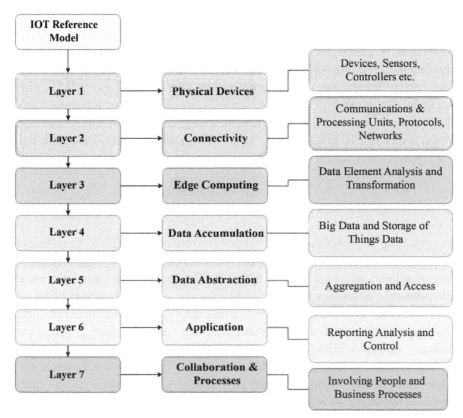

FIGURE 5.1 Layers of IoT architectural model.

Layer 1 is called the physical layer. The data is collected from the surroundings through hardware and it is transferred to the upper layers. Sensors and devices usually form this layer, which are used to collect data such as pressure, temperature, water quality, humidity, air quality, and so forth, (Alenezi et al., 2017) from the environment (Chain of Things, 2017). Layer 2 is called the connectivity layer, which connects various devices in the IoT network with each other with the help of interconnection devices such as routers, switches, and gateways. It is responsible for transferring the data collected from sensors securely to various upper layers for processing of data. Layer 3, called edge computing layer, is responsible for analysis of data obtained and storing this information for higher-level processing. Data transformation techniques are used, which reduces the size or amount of data flowing. This information is accumulated in layer 4, which changes the event-based information to query-based data for upper layers. Abstraction of data takes place in layer 5 where the data flowing from various sources is combined and converted into storage format appropriate for applications.

Layer 6 is called the application layer, which functions to interpret information for various IoT-based applications, which includes smart homes, cities, grids, healthcare, agriculture, and so forth (Stallings, 2015). The final layer, that is, layer 7, identifies various people who can integrate their resources to utilize the data obtained efficiently. It also includes creating visualizations and business models from information extracted through the application layer (Das, 2015).

5.4   DIFFERENCE BETWEEN TRADITIONAL NETWORKS AND IOT

It is important to understand how an IoT infrastructure differs from traditional networks for the development of required privacy, security, and resource-related solutions for IoT systems. The most significant difference scales down to resources level available for the devices involved (Jing et al. 2014). Since IoT usually involves presence of embedded devices such as sensors, RFID equipment, BLE devices, and so forth, there is always a constraint on the available resources. They usually require small battery life, low computing power as well as memory, which are not a concern for traditional networks. (Christidis and Devetsikiotis, 2016). Due to this fact, traditional networks such as powerful computers, smart phones, and servers can be secured through multifactor authentication as well as complex security protocols. Whereas, IoT systems require lightweight and not resource-heavy security protocols that maintain a fair balance between resource consumption and security.

IoT devices primarily use low power and bandwidth-efficient medium such as 802.11a/b/g/n/p, ZigBee, SigFox, and LoRa for connecting to the internet or the gateway. In contrast, traditional networks use wired/wireless medium such as DSL/ADSL, WiFi, 4G, and fiber optics for communication. These are more secure and fast. IoT devices also lack OS in most cases due to their application-specific functionality whereas traditional networks share similar OS and data formats. This diversity challenges the development of a standard security protocol that can be used by all types of IoT devices.

Security design for traditional networks is way different from that of IoT as they are secured by static network perimeter defense mechanisms like firewalls. The end devices are also protected by host-based applications such as antivirus, anti-malware, software patches, and so forth, which cannot be applied to IoT because of the resource limitation (Yu et al., 2015). The absence of host-based defense measures, physical security layer, access control measures, and security patches makes IoT systems vulnerable to insider attacks. Thus, IoT is considered as a perfect target for the attackers due to its low resource requirement and less secure communication protocols. This contributes to a lack of standardization in IoT solutions around the world.

5.5   IOT THREATS AND CHALLENGES

The various challenges in the development of IoT infrastructures based on resource constraints are described in this section.

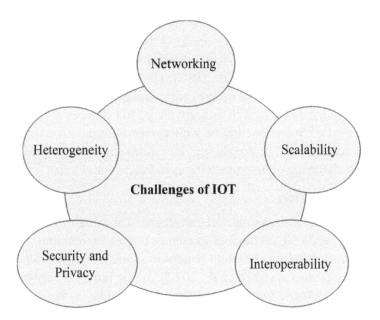

FIGURE 5.2   Threats related to IoT.

5.6   SECURITY AND PRIVACY THREATS

The potential threats and vulnerabilities to IoT systems will increase by 2020 (Ahlmeyer and Chircu, 2016) with the exponential rise in connected devices. To make the situation even worse, complicated cyberattacks like

Ransomware (Brewer, 2016), Mirai, and DuQu-2 have been successfully launched, which have made the existing IoT protocols ineffective. This proves that IoT systems are quite vulnerable to ransom payment, data forgery, data theft, and suspicious botnet attacks. Moreover, despite controlled access to data and centralized network, the cloud-supported IoT has privacy and security issues (Puthal, 2016). A typical cloud service is vulnerable to single point of failure and the data privacy can be breached causing unauthorized data sharing (Kshetri, 2017). The issues related to IoT security can be related to scarcity of resources and poor design of security modules. Such security flaws have led to attacks on data integrity, device integrity, and privacy as well as attacks on network such as DDoS (Distributed Denial of Service) (Jing et al., 2014).

Heterogeneity

IoT involves billions of different devices that makes heterogeneity a big issue. The main aim of such IoT systems is to build a common platform to abstract heterogeneity of the devices involved to maximize the exploitation of their functionality (Alenezi et al., 2017). Thus, with the significant growth in IoT platforms, development of applications that can adapt themselves to the variation in hardware as well as software of IoT devices remains a challenge. The manufacturers need to be well versed with the diversity existing in network connectivity, protocols, and communication techniques in order to tackle the challenge of heterogeneity in IoT (Khan and Salah, 2018).

Adaptive networking

A suitable network is the primary requirement of the IoT environment that drives and connects all the devices to ensure proper functioning. However, these devices are present in different structures, which require different types of networking protocols (Chen et al., 2012). While building network architecture for IoT systems, appropriate protocols should be chosen considering the impact on system's performance as well as usability.

Interoperability

The capability of components to communicate and cooperate with each other to efficiently produce desired results despite technical differences is called Interoperability. As the number of heterogeneous devices increase in an interconnected network of IoT, the interoperability becomes a priority to ensure efficient functioning of the entire network. Interoperability remains a challenge in the creation of a typical IoT network (Bubley, 2017).

Scalability

Scalability of a system is defined as the ability to function efficiently without any loss in performance with the increase in the number of connected devices. With variation in functions performed by IoT systems, they ought to be scalable for proper functioning. Therefore, scalability remains a challenge for IoT (Gupta et al., 2017).

5.7   SECURITY REQUIREMENTS OF IOT

Standardized form of security requirements for a typical IoT system is described in Figure 5.3. IoT system should be capable of carefully functioning in an environment with less trust factor. Since sensor network is an integral part of IoT, there is a need for security against unauthorized data manipulation and sharing. Moreover, because most IoT systems are being deployed in public areas, it is also required to provide physical protection, that is, against damage also. The integrity of code as well as data being recorded should be ensured (Sadeghi, 2015).

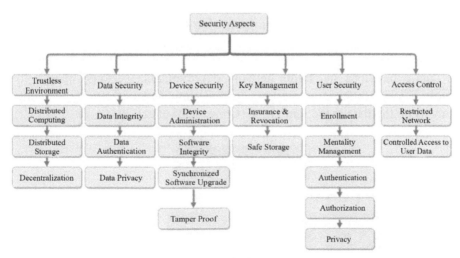

FIGURE 5.3   Security requirements for a typical IoT system.

Security requirements for a typical IOT system are shown in Figure 5.3.

The network being employed should be immune to malware attacks, and the devices should be properly authenticated and authorized before adding them as a network node. It is required to take care of hardware- as well as software-wise tampering of devices. Keeping in mind the aspect of resource management, direct cryptographic measures cannot be easily implemented in IoT systems as memory, power, and computation is limited (MalviyaH, 2016). This calls for development of lightweight cryptographic measures with an efficient system to manage specific keys. Along with security measures for infrastructure, client-side privacy, and security is also required. Unique client ID management, authentication, authorization, and registration are the key aspects. Thus, there is a need for a standard global protocol ensuring authentication in IoT systems.

5.8   PERFORMANCE REQUIREMENTS AND ENERGY EFFICIENCY

The sharing of data occurs in real time in communication network of the IoT system, the performance efficiency is really important. Figure 5.4 depicts various requirements when it comes to performance of a designed IoT framework. IoT frameworks should be self-manageable and self-regulated to protect them from human errors. An IoT system is termed as efficient if it smartly caters to the constraint in resources of end devices as well as network including low-power consumption, computational power, and memory. This should not come at the cost of compromising the security. With increase in the number of users and devices within the framework, more data will be generated. Therefore, IoT system should be flexible enough to accommodate for data expansions and future network deployments. It should be able to handle a large number of requests in real time and perform appropriate analysis on the data at the same time (Conoscenti et al., 2016).

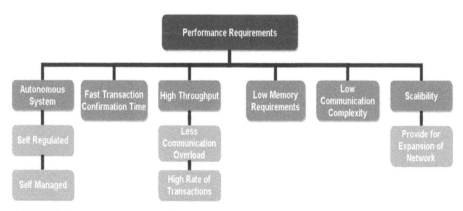

FIGURE 5.4   Performance requirements of a typical IoT system.

There is a need of sophisticated security mechanism for IoT. Thus, many researchers visualize blockchain as the key solution to all the problems discussed above. Therefore, following sections will describe aspects of blockchain technology that can be used to integrate it with IoT systems.

5.9   CHARACTERISTICS AND KEY ASPECTS OF BLOCKCHAIN

Due to the decentralized and secure architecture of blockchain, it ensures the integrity and trust of transactions being performed. This section discusses various key aspects of blockchain framework that can prove advantageous for integration with IoT systems. These characteristics are summarized in Figure 5.5 and are discussed next.

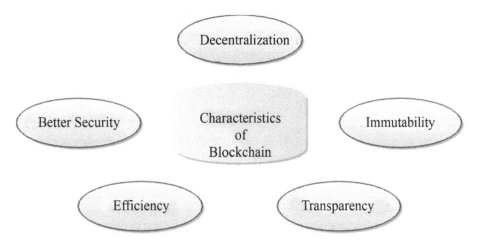

FIGURE 5.5   Blockchain characteristics.

Decentralization

The most important property of any blockchain network is its decentralized architecture, which prevents a lot of issues such as scalability and single point of failure. The processing capabilities of users involved in blockchain network are utilized efficiently with the help of a distributed ledger thus reducing the problems of latency and eliminating the single point of failure (Gupta et al., 2017).

Immune ledgers

The integrity of data can be maintained by creating immutable ledgers of transactions performed. In typical architectures, the databases can be easily changed and they require a trust with outside party to maintain the integrity of information. This is not the case with blockchain as every single block of the network is related to the block preceding it by forming a chain of blocks that cannot be altered as far as the user remains in the network (Sikorski, 2017).

Improved security and privacy

Blockchain has an advantage when it comes to protection against malicious activities on important data as it uses public key infrastructure. The users involve in the network share their trust in the security features and uses consensus mechanism. These users form anonymous entities of the network, which can create and propose a transaction, validate them, and establish the integrity of the data. When these transactions are created, they are signed by various nodes requiring unique private key to validate their true ownership of the particular asset in use, which is further transferred to the next node in the network.

Encryption

Blockchains primarily use hashing techniques to secure their transactions. In a typical blockchain infrastructure, each transaction uses a block that is hashed to generate a hash value. In this way, the data is bound with very strong cryptographic methods that cannot be tampered or tracked by unauthorized users. Once these transactions are validated and verified by consensus, the data present in the block becomes immune to alterations (Prisco and Slock, 2016).

Improved efficiency

The typical centralized architecture is improvised in blockchain by distribution of database records among users involved within the network. Thus, transparency in verification of all records stored in the database is increased by distribution of transactions. Christidis and Devetsikiotis show that a typical centralized framework is less efficient than a blockchain architecture in terms of risk management, cost, and speed of transaction settlement.

Integration of blockchain and IoT

IoT has become an integral part of the digital era with continuous transformation and optimization of manual tasks by analyzing volumes of data and information at different levels. The quality of life has been improved with development of smart applications that are facilitated by these information as well as digitization of services. Many platforms to manage and access such voluminous information have also surfaced due to unprecedented growth of IoT systems. This exponential growth has led to many vulnerabilities being developed due to the lack of confidence of centralized architectures like cloud computing, which forms a significant part of any typical IoT infrastructure. Therefore, integrating cloud computing with IoT has advantages, which come at a cost of security and privacy. As discussed earlier, the revolutionizing potential of blockchain technology can be integrated with IoT to achieve efficient results. IoT platform can be enriched through blockchain, which brings trust as they are reliable and traceable. The data remains immutable for the entire time and the sources can be easily identified as well. Blockchain can transform IoT applications where information is required to be shared securely between large groups of participants. Such applications may include healthcare, manufacturing, smart appliances, cities, food care, and so forth. The list is endless. In applications such as smart cars and cities, secured sharing network could encourage better participation within the developed ecosystems thus improvising the quality of life at a greater scale. It can be seen that use of blockchain can complement the IoT industry and it has started gaining recognition as well (Malviya, 2016).

Advantages of blockchain–IoT integration

Blockchain is expected to solve a variety of challenges faced by traditional IoT networks. We have summarized some of the improvisations in this section.

The distributed P2P architecture of blockchain promotes decentralization of power that prevents various scenarios where powerful companies have full control over information extracted from a huge number of people. It eliminates the threat of single points of failures thus minimizing the bottlenecks in operation of IoT systems. This further improves the tolerance of architecture against faults thus boosting system scalability. An important property of blockchain is the identification of every single device of network participants. The data fed into the system can be uniquely identified through the actual data that is immutable. This property of blockchain can ensure authorization of devices used in IoT as well as authentication of trusted distributers (Gan, 2017). The development of smart autonomous applications is fully supported and empowered using blockchain technology as it contains latest application features. Blockchain can facilitate interaction of such devices with each other without involvement of servers as middleware.

Blockchain can ensure immutability of information extracted through IoT devices thus improvising system reliability (Amoozadeh et al., 2015). Devices present in an IoT network can be authenticated with this mechanism and thus making them tamper proof. Data extracted through sensors can be traced using blockchain technology. Blockchain thus brings the key aspect of reliability to the IoT framework. (Ho et al., 2017).

The information flowing through the IoT network and communications occurring between components of the framework can be secured by storing them as transactions of a typical blockchain network (Prisco and Slock, 2016). The messages being exchanged between devices can be treated as transactions by the blockchain, which are further validated as smart contracts thus ensuring secure communication between devices. Integration of block-chain can optimize the current security protocols applied to IoT frameworks (Khan and Salah, 2018). Blockchain can secure the source code through immutable storage mechanism and can facilitate secure installation in the devices of IoT network. Manufacturers of such devices can securely update and track with high levels of privacy. Middleware services present in IoT can benefit from this functionality by securely updating IoT devices.

The creation of IoT infrastructure can be accelerated through blockchain in places where transactions does not require authentication from authorities. It can facilitate micro payments and deploy micro services safely in a trust less environment. The access of IoT data can be improvised in blockchain. Thus, it can be seen that integration between IoT and blockchain can address the limitations of IoT systems while maintaining the data reliability, security, and privacy.

5.10   IMPROVING THE SECURITY OF AN IOT SYSTEM USING BLOCKCHAIN

For exchanging the data with each other, IoT devices process a transaction and store the result into a ledger. Encryption keys can be stored in these ledgers making the exchange of information more confidential. The message sent by IoT device can be encrypted using the public key generated by the device at end point. This key should be stored in the blockchain network. The device sending the data asks its respective node to receive the public key from receiver point through the ledger. This message is encrypted by the sender device using public key of receiving device, which makes sure that only the receiving device can decrypt and read the message using its private key. The message being sent is digitally signed by the sender device, which is verified by the receiver using public key extracted from the ledger. This authentication process starts by calculating the hash of a particular message that is encrypted using a private key. The signature and message are transmitted by the sender into the communication channel where receiver receives and decrypts the signature using public key of sender stored in the ledger to obtain the hash value calculated earlier. The calculated value of hash is compared to the original hash value of the message, which remains valid only if both are same. This improvises the trust of received messages as each message is stored in the ledger of the blockchain (Das, 2015).

In a typical IoT network, a large number of devices are connected to the same network, which makes it necessary for the system to differentiate the infected nodes from legitimate ones (Das and Maniklal, 2015). The exchange of information in a device can only occur once it has registered itself in a node of IoT network. It can thus receive and send information from other peers on the network using numerous nodes present. Thus, for a private network, only legitimate devices could be added to the blockchain network. New devices are required to be authenticated by root servers before appending it to the node list. For the same, a set of credentials are generated as a part of blockchain implementation on such devices during the setup stage. This maintains the integrity of the network as a device is enrolled into the secured network only if the credentials are authenticated by the root (Alenezi et al., 2017).

IoT devices can be installed using secure configurations through blockchain technology by hosting information such as configuration details and firmware updates on the ledger. The configuration is encrypted thus the topology of the IoT network is secured. Also, the hash value of configuration files can be stored in the ledger. The device present at the node of blockchain can compare the hash value and configuration can thus be installed to the device by the administrator. The boot controls can also be secured by the same method.

5.11   EXISTING BLOCKCHAIN—IOT PLATFORMS

Numerous blockchain platforms have emerged, which are focused on IoT as the industry grows in popularity. Some of them are listed here.

- IOTA: IOTA is one of the first blockchain platform for IoT, which was designed specifically for IoT applications and it provides a transaction settlement and data transfer layer for devices connected to the IoT network (Open-Source Distributed Ledger). They have created a block less, cryptographic, and decentralized platform called Tangle (Brewer, 2016) which verifies transactions of every single user instead of outsourcing network verification.

- HDAC: The Hyundai Digital Asset Company (HDAC) has applied blockchain technology for effective communication between IoT devices whereby verification of identity, its authentication and storage of data, are all handled at the same time. They achieved increased transaction rates as well as volumes by incorporating a double-chain system, both public as well as private, ideally for IoT devices (Bubley, 2017). Smart homes, factories, and buildings are their major domains of operations (Bubley, 2017).
- VeChain: It is a public blockchain platform that is developed at a global enterprise level. Their primary focus is to integrate IoT in cold chain logistics through proprietary IoT devices for tracking the key metrics like temperature and humidity throughout the journey. The platform is successful in holding automobile passports through digital records of cars including information such as insurance, repair history, registration along with the behavior of driver throughout the journey (Chain of Things, 2017). They have also expanded to medical and healthcare domain by using end-to-end tracking of product process of medical devices.
- Waltonchain: A combination of RFID and blockchain technologies is used to create Waltonchain for effective IoT integration. Their aim is to track processes and products present in the supply chain, where technology can be applied for high-end clothing identification, tracing food and drugs, and logistics by implanting RFID tags and reader–writer control chips into the products (Chen, 2012). The information regarding status of involved products is collected and analyzed onto a secure blockchain.

5.12   IOT BLOCKCHAIN INTEGRATION USE-CASE

We consider an example of a self-driving car, which is equipped with number of sensors and actuators, sophisticated algorithms, and powerful processors to execute software. The sensors in smart cars are mainly used for (1) navigation and guidance, (2) controls and safety, and (3) performance monitoring. Sensors like GPS module, radar systems, engine temperature sensor (Huh et al., 2017). Air temperature sensor, fuel temperature sensor, cabin temperature sensor, vehicle-to-vehicle (V2V) and vehicle-to-infrastructure (V2I) technology (to allow the car to communicate with other cars and the infrastructure like traffic lights), and so forth. The model consists of three layers in the network, the self-driving car, an overlay network, and the cloud storage.

The V2V and V2I technology allows cars to interact it with other vehicles and traffic lights. Vehicle speed and gap between vehicles can be controlled in response to the conditions on the road. The communication between vehicles needs to be secured as any kind of interruption or manipulation in the data can cause serious road accidents. Blockchain is adopted for securing the data transactions. Considering data storage and access use cases, the devices should be able to store the data on storages to be used by a third party to avail services and for research (Ducklin, 2016).

5.13   MODEL ARCHITECTURE

The three layers in the network, that is, the self-driving car, an overlay network, and the cloud storage are discussed herein.

Smart car

The smart car includes: the smart devices (sensors), a local blockchain, and a local storage. The smart devices collect all the data from the smart car and send the data to the blockchain. The local blockchain can be mined and is stored on a device that is always online, such as a smart hub. A starting transaction is created to add new devices to the policy header of the local blockchain, which is the access control list for blockchain. This can be done by providing them a shared key based on a generalized algorithm. The local storage is an optional storage device, used for creating backup of the data (Stallings, 2015).

FIGURE 5.6   Smart car demonstration.

Overlay network

The overlay network is a peer-to-peer decentralized network in the model. The nodes in the network could be the devices in the smart cars, the smart car miner (like smart hub) or third-party users from the network that have access. There can be multiple nodes in the overlay network relating to same device so to decrease any kind of delay, the nodes in the overlay network are grouped together and a pointer for the group is chosen to point to the group of nodes. The group head contains the list of keys that can access the data from the nodes, the list of keys of other vehicles that can be accessed and a list of data transactions to other groups in the network. Every pointer in the overlay network keeps a record of the data transactions by cloud storage and access control as blocks in the blockchain. Every new block is mined and a decision is made whether to keep the new block in the chain or not based on its communication with transaction participants. If found appropriate, it is added to the blockchain and the miner adds the policy from previous block to the new block. This decision-making leads to variation in the blockchain at different group pointers.

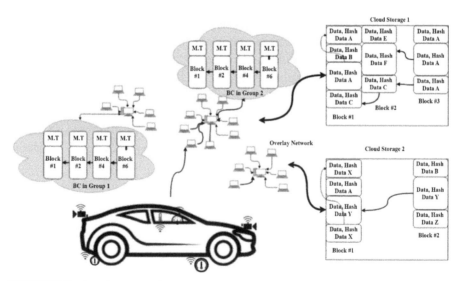

FIGURE 5.7   Overlay network.

Cloud storage

The data from the devices in self-driving cars is stored in cloud storage, so that the decision-making for the control systems of the vehicle can be improved. This increases the efficiency of the vehicle and decreases any kind of damages to vehicle (Sikorski et al., 2017). The data can also be shared with third parties who may provide smart solutions like temperature adjustment and service management, optimized insurance premium generation, and many others. The cloud stores the data from a single vehicle in similar blocks with a unique key for the vehicle. The key for the blocks is encrypted by a generalized algorithm to ensure the security of data. The key is used for authenticating the access to data in the associated blocks.

Data transactions

The section discusses the transactions control in the presented blockchain–IoT model.

Storage

The data from devices is stored in the cloud and a backup device is connected locally to save data, a shared storage can also be used to serve this purpose. To store data in a cloud storage system, an account is required with the cloud service provider and then permissions are given to the devices to send data to the storage. The cloud provides the device with a pointer to first data block and when the device needs to send data to the storage, it uses this pointer to send data. The devices send the data to the miner, the miner fetches previous block pointer and hashes the data with a key and sends the data to the storage with the key. When the cloud facility receives the data, a verification of data is performed by hashing the data and matching it with the key sent by the miner. If the hash matches the key, the data from the device is stored in the cloud and the block pointer is encrypted with an algorithm to generate a key. The hashed data is forwarded in the overlay network and is mined in the blockchain to make any kind of changes to make the data public. (Brewer, 2016).

While using local storage, the data is not required to be hashed and no verification is needed since all the communication and transactions are performed in the local system and without any interference from outside.

Access

The data from different devices will be required to be accessed for evaluation time to time in order to implement new services and improve on previous versions. To access the data, the requester and requestee needs to sign a transaction as created by the service provider. The request is sent to the pointer of the group associated with the device. The group pointer checks the list of permitted keys and if it finds a valid key in the list, it proceeds with the transaction from its group; else, the pointer broadcasts the transaction request to another pointer. If the key is available in the list, the miner requests the data from storage. It then encrypts this data with the requesters' key and sends it in response. Additional security can be implemented by the miner through the introduction of noise in the data. After sending data, the miner stores the transaction in the local storage and sends the transaction to random groups for storing the transaction, which is a proof that the data was sent and can be used to trace back any kind of misbehavior. (Shah and Yaqoob, 2016).

If the requester is authorized to access the entire chain of data and a request for the same is received, the miner sends the pointer and the hash of data in the storage. In other cases, the miner tries to minimize the data being sent to the third party by adding safe answer to data.

Monitoring

In some cases, the vehicles may be monitored in real time to obtain certain information. A monitoring transaction is created for this purpose. This transaction makes a connection between the requestor and requestee, the miner requests data from the device/s and forward the data to the requestee. The transaction is stored in blockchain as one block with time stamps and duration of monitoring.

5.14   OPEN CHALLENGES AND FUTURE PROSPECTS

The different challenges for efficient implementation of security for IoT devices are discussed (Jing et al., 2014).

Limitation of resources

One of the main obstacles in describing a security mechanism that is sturdy and robust is resource–regulation architecture of IoT. As compared to the conventional standards, the cryptographic algorithms have a restricted job of working within these limitations. The requirement for exchange of keys and certificates with any broadcasts and multicasts, the storage and the energy requirements need to be overcome for a successful implementation of security and communication protocols for IoT. It includes remodeling of the given protocols in order for them to be lightweight and energy-saving although requiring complex computations accompanied by improvement of energy harvesting techniques (Kamalinejad, 2015).

Heterogeneity

Because of the presence of varied devices ranging from small, low power-driven objects with sensors to high-end servers, a more sophisticated high-dimensional security framework is required. Initially, the framework needs to adjust itself to deployed resources, make judgments for selecting the security procedures at IoT layers after that any facility or resource can be given to end users. Such a robust and easily adjustable security structure demands intelligence, which requires standardization of resources to be positioned in IoT architectures (Amoozadeh et al., 2015).

Regarding security

For having a standard global security procedure for IoT, conversion mechanisms should be provided to the protocols that are implemented at different levels. Considering the global process, a combination of security benchmark at every layer can then be defined effectively taking into account architectural constraints. (Malviya, 2015).

5.15   SINGLE POINTS OF FAILURE (SPOF)

Due to the diversified networks, architectures and protocols, the IoT system has become more susceptible to single points of failure as compared to any other system. A single point of failure (SPOF) if occurs then it may terminate the working of entire system. A lot of research work is yet to be done to make sure sufficient availability of IoT elements, especially for mission-critical applications. All this will demand techniques and standards in order to ensure redundancy. This requires trade-off between the reliability and costs of the entire infrastructure.

5.16   SUSCEPTIBILITIES TO HARDWARE OR FIRMWARE

The architecture of IoT has become more susceptible to vulnerabilities of hardware with the global growth of devices that are of low-cost and low-power driven. Not only the physical malfunctioning, but it is required to manage security algorithms in the hardware, routing, and packet managing procedures also need to be validated before being used in IoT as any fault found after deployment will be difficult to rectify. Therefore, a robust verification protocol is a must for capitalizing IoT security.

Assured updates and management

One of the main issues for future research is to make available millions of IoT devices, trusted and powerful management and updates of software. Furthermore, the issues regarding secure and trusted governance of IoT device ownerships, data privacy, and supply chain are research problems that need the focus of the researchers in order to promote a widespread adoption of IoT. The blockchain technology can be considered as the facilitator of IoT security solutions, but the technology still has challenges and hurdles to overcome, which includes its efficiency, key collision, and regulations.

Vulnerabilities of blockchain

Although providing with sturdy approaches for a safe and secured IoT, the blockchain systems are still considered to be susceptible to many challenges (Li et al., 2017). The different procedures depending on the hashing power of miner can be easily compromised, which allows the hacker to host the blockchain. In the same way, limited randomness of the private keys can be very well exploited to compromise the blockchain accounts. Hence, effective systems are required to be defined and implemented so as to ensure the privacy of transactions. The race attacks need to be avoided, which may cause extra spending during transactions (Cisco. The Internet of Things Reference Model. White Paper, 2014).

5.17   CONCLUSION

Advanced technologies such as IoT are termed as disruptive as it has created a great controversy. The security and privacy issues in IoT can be addressed by the revolutionizing technology of blockchain. It can provide an undeniable support in the long-term technological investment and production. Blockchain can be integrated with IoT with caution in order to prevent the risks involved with it. This chapter has provided an analysis of various advantages, key challenges, and difficulties in the way of blockchain and IoT integration. We have listed down the key points where blockchain technology can help improvise and secure various IoT applications. Various existing platforms and applications were examined to support our study by offering an overview of how blockchain interacts with an IoT ecosystem. It is the need of the hour to design and develop a secure blockchain-based IoT system that meets various requirements of power as well as privacy to give a boost to the development of an autonomous digital world. These developed ecosystems should be compatible with the current IoT infrastructure so that proper and self-sustained transformation can take place into the decentralized architecture that is economically feasible (Khan, and Salah, 2018).

KEYWORDS

- **Internet of things**
- **security**
- **reliability**
- **blockchain**
- **blockchain-based IoT**

REFERENCES

Ahlmeyer, M.; Chircu, A. M. Securing the Internet of Things: A Review. *Issues Inf. Sys.* **2016,** *17* (4).

Alenezi, A.; Zulkipli, N. H. N.; Atlam, H. F.; Walters, R. J.; Wills, G. B. The Impact of Cloud Forensic Readiness on Security. In *International Conference on Cloud Computing and Services Science*; Scitepress: 2017; Vol. 2, pp. 539–545.

Amoozadeh, M.; Raghuramu, A.; Chuah, C. N.; Ghosal, D., Zhang; H. M., Rowe, J.; Levitt, K. Security Vulnerabilities of Connected Vehicle Streams and Their Impact on Cooperative Driving. *IEEE Commun. Mag.*, **2015,** *53* (6), 126–132.

An Open-Source Distributed Ledger. https://www.iota.org/get-started/what-is-iota.

Brewer, R. Ransomware Attacks: Detection, Prevention and Cure. *Network Security* **2016,** *2016* (9), 5–9.

Bubley, D. Data over Sound Technology: Device-to-Device Communications and Pairing without Wireless Radio Networks. *Int. J. Comp. Intell. Res.* 2017.

Chain of Things. (2017). Available online: https://www.blockchainofthings.com/.

Chen, M.; Wan, J.; Li, F. Machine-to-Machine Communications: Architectures, Standards and Applications. *Ksii T. Internet Inf.* 2012, *6* (2).

Christidis, K.; Devetsikiotis, M. Blockchains and Smart Contracts for the Internet of Things. *IEEEAccess* **2016,** *4*, 2292–2303.

Cisco. The Internet of Things Reference Model. White Paper, 2014.

Conoscenti, M.; Vetro, A.; Martin, J. C. D. In *Blockchain for The Internet of Things: A Systematic Literature Review*. Proceedings of the IEEE/ACS 13th International Conference of Computer Systems and Applications (AICCSA), 2016.

Das, M. L. Privacy and Security Challenges in Internet of Things. In *International Conference on Distributed Computing and Internet Technology*; Springer: Cham, 2015; pp. 33–48.

Ducklin, P.; Mirai. Internet of Things. Malware from Kreb DDoS Attack Goes Open Source. https://nakedsecurity.sophos.com/2016/10/05/mirai/.

Duqu 2.0: The Most Sophisticated Malware Ever Seen (Updated 2018). (2019). Retrieved from http://resources.infosecinstitute.com/duqu-2-0-the-most-sophisticated-malware-ever-seen/#gref.

Gan, S. An IoT Simulator in NS3 and a Key-Based Authentication Architecture for IoT Devices using Blockchain. Indian Institute of Technology: Kanpur, 2017.

Gupta, A.; Christian, R.; Manjula, R. Scalability in Internet of Things: Features, Techniques and Research Challenges. *Int. J. Comput. Intel. Res* **2017,** *13* (17), 1617–1627.

Ho, G.; Leung, D.; Mishra, P.; Hosseini, A.; Song, D.; Wagner, D. In *Smart locks: Lessons for Securing Commodity Internet of Things Devices.* Proceedings of the 11th ACM on Asia Conference on Computer and Communications Security; ACM: New York, 2016; pp. 461–472.

https://bitcoinmagazine.com/articles/slock-it-to-introduce-smart-locks-linked-to-smart-ethereumcontracts-decentralize-the-sharing-economy-1446746719/.

https://coincentral.com/waltonchain-beginner-guide/.

https://cryptobriefing.com/what-is-vechain-introduction-to-vet-thor/.

Huh, S.; Cho, S.; Kim, S. Managing IoT Devices using Blockchain Platform. In *2017 19th International Conference on Advanced Communication Technology (ICACT)*; IEEE: New Jersey, 2017; pp. 464–467.

Hyundai Digital Asset Currency (HDAC) for a Secure Internet of Things (IoT). https://medium.com/amazix/hyundai-digital-asset-currency-hdac-for-a-secure-internet-of-things-iot-c4038a556977.

Jing, Q.; Vasilakos, A. V.; Wan, J.; Lu, J.; Qiu, D. Security of the Internet of Things: Perspectives and Challenges. *Wireless Networks* **2014,** *20* (8), 2481–2501.

Kamalinejad, P.; Mahapatra, C.; Sheng, Z.; Mirabbasi, S.; Leung, V. C.; Guan, Y. L. Wireless Energy Harvesting for the Internet of Things. *IEEE Commun. Mag.* **2015,** *53* (6), 102–108.

Khan, M. A.; Salah, K. IoT Security: Review, Blockchain Solutions, and Open Challenges. *Future Gener. Comp. Sys.* **2018,** *82*, 395–411.

Kshetri, N. (). Can Blockchain Strengthen the Internet of Things? *IT Professional* **2017,** *19* (4), 68–72.

Li, X.; Jiang, P.; Chen, T.; Luo, X.; Wen, Q. A Survey on the Security of Blockchain Systems. *Future Gener. Comp. Sys.* 2017.

Lund, D.; MacGillivray, C.; Turner, V.; Morales, M. Worldwide and Regional Internet of Things (IoT). Forecast: A Virtuous Circle of Proven Value and Demand. *International Data Corporation (IDC)* 2014.

Malviya, H. How Blockchain will Defend IOT. Available online: https://ssrn.com/abstract=2883711. (2016).

Prisco, G.; Slock.it to Introduce Smart Locks Linked to Smart EthereumContracts, Decentralize the Sharing Economy. 2016.

Puthal, D.; Nepal, S.; Ranjan, R.; Chen, J. Threats to Networking Cloud and Edge Datacenters in the Internet of Things. *IEEE Cloud Computing* **2016,** *3* (3), 64–71.

Sadeghi, A. R.; Wachsmann, C.; Waidner, M. Security and Privacy Challenges in Industrial Internet of Things. In *2015 52nd ACM/EDAC/IEEE Design Automation Conference (DAC)*; IEEE: 2015; pp. 1–6.

Shah, S. H.; Yaqoob, I. A Survey: Internet of Things (IOT) Technologies, Applications and Challenges. In *2016 IEEE Smart Energy Grid Engineering (SEGE)*; IEEE: 2016; pp. 381–385.

Sikorski, J. J.; Haughton, J.; Kraft, M. Blockchain Technology in the Chemical Industry: Machine-to-Machine Electricity Market. *Appl. Energy* **2017,** *195,* 234–246.

Stallings, W. The Internet of Things: Network and Security Architecture. *Internet Protoc. J.* **2015,** *18* (4), 2–24.

The Tangle: An Illustrated Introduction. https://blog.iota.org/the-tangle-an-illustrated-introduction-4d5eae6fe8d4.

What is VeChain? Introduction to VET (THOR).

What is Waltonchain (WTC)? A Guide to the IoT and Blockchain.

Wurm, J.; Hoang, K.; Arias, O.; Sadeghi, A. R.; Jin, Y. (, January). Security Analysis on Consumer and Industrial IoT Devices. In *2016 21st Asia and South Pacific Design Automation Conference (ASP-DAC)*; IEEE: 2016; pp. 519–524.

Yu, T.; Sekar, V.; Seshan, S.; Agarwal, Y.; Xu, C. In *Handling a Trillion (Unfixable) Flaws on a Billion Devices: Rethinking Network Security for the Internet-of-Things*. Proceedings of the 14th ACM Workshop on Hot Topics in Networks; ACM: 2015; p. 5.

Blockchain Disruptive Use Cases, 2016. Available online: https://everisnext.com/2016/05/31/blockchain-disruptive-use-cases/.

CHAPTER 6

Securing IoT with Blockchains

RASHMI AGRAWAL* and SIMRAN KAUR JOLLY

MRIIRS, Faridabad, Haryana, India

Corresponding author. E-mail: drrashmiagrawal78@gmail.com

ABSTRACT

Blockchain technology was developed in 2008 as the next evolutionary technology. Blockchains, basically are sequences of blocks connected via cryptography which is resistant to change in the data. One of the important applications of blockchain is providing security to all the IOT devices which we have discussed in the chapter. All the devices that are connected via the internet may synthesize with each other so their security is the main concern here. So, the devices can be made secure by using RSA (Rivest–Shamir–Adleman) cryptosystems that ensure security via a private key in all the devices. Hence, we present the RSA-based security in our IOT (internet of things) devices. In our study, we review the blockchain technology and its applications in maintaining the security in fully scaled IOT systems using a simple key management system. One of the most quintessential components of blockchain technology is its ability to act as a trusted ledger of transactions so that it does not have to rely on other client–server architectures. The important modules of blockchain are (1) registration, (2) authentication, and (3) check listing IOT devices.

6.1 INTRODUCTION

Blockchain technology is a decentralized system of records of transactions which can be modified anytime without changing the records. Blockchain comprises two words block and chain, which means it is a sequence of blocks that share some certificates (data). This technology is used for secure communication between two parties in that particular block. This technology was invented by Satoshi Nakamoto in 2008 (Nakamoto, 2009). The framework of blockchain technology encompasses the following components:

a. **Blocks:** These are the building blocks of blockchain technology to store the transactions in a cryptographic hash function. A cryptographic hash function takes input message of any length and produces the hash function. The two blocks are linked via a hash function forming a chain. Whenever a new transaction is added in the database it is updated in the new block. The previous blocks are removed.

b. **Timestamps:** Every block in a blockchain has a timestamp associated with it. A shorter block time means faster transactions.

c. **Decentralization:** A blockchain system is decentralized in nature which means all the nodes are generated in the form of a tree. There is no central point of failure in the system.

d. **Private blockchain:** These blockchains need protection from external attacks as they cannot rely on any anonymous factors.

6.2 APPLICATIONS

1. Crypto-currencies: These are used to record transactions in the bitcoin technology (Abduvaliyev et al., 2013).
2. Banks: Banks are using smart contracts to provide security inside their systems without using any third party for transactions between two parties.
3. Monitoring: Big retail stores like Wal-Mart are using blockchain technology to monitor the supplies that are coming out and going in.
4. Security: Blockchains are also being used for providing security in IOT devices to ensure secure communication between two devices.
5. Healthcare: All the health records of the patients are stored on blockchain with a private key. This technology could be used for health management, storing supplies of drugs, and managing them efficiently.

6.3 TWO TYPES OF BLOCKCHAIN

There are two types of blockchain:

Public Blockchain

Everyone can read or modify data in this type of blockchain. In the public blockchain network, anyone can join the network and it is one of the largest networks in blockchain technology. The only limitation of this mechanism is the high processing power that is needed to maintain so many transactions. As everyone participates without any access control so privacy is an issue in the public blockchain network.

Private Blockchain

In this type of blockchain all the parties are trusted between which exchange of data occurs. This type of blockchain occurs when the parties belong to the same entity. These are controlled blockchains which give permission to modify and read the data in the blockchain network. Hence, access control mechanisms are used in these networks to give permission to read and write on the network. Figure 6.1 further explains the portioning of blockchain into four subparts and their application areas (Stanciu, 2017).

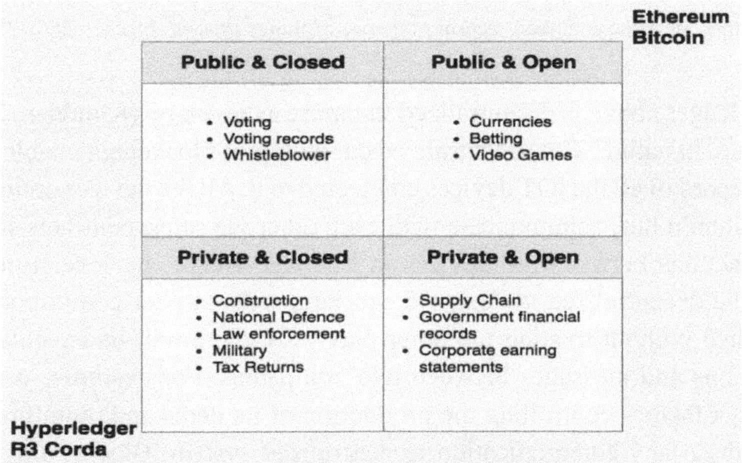

FIGURE 6.1 Classification of blockchain.

6.4 BLOCKCHAIN AND IOT

Blockchain is the vital element of IOT industry (Stanciu, 2017) which is used to connect a large number of devices, maintaining scalability in the system, allowing transactions, and data privacy. Hence the cryptographic algorithms used in blockchain are used to maintain data privacy. Hence blockchain is a ledger of bitcoin used to maintain security in the network of devices. Figure 6.2 shows the function of a blockchain as a ledger between different transactions.

The blockchain functions as a distributed transaction ledger for various IoT transactions

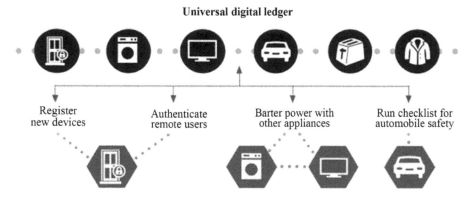

FIGURE 6.2   Blockchain as a ledger.

Source: https://datafloq.com/read/securing-internet-of-things-iot-with-blockchain/2228

The ledger above is decentralized in nature as given by (Khudnev, 2017) and secure in nature. The decentralized capability of blockchain enables it to keep a record of all the IOT devices connected to it. All the devices connected via a common bus, communicate with each other via smart contracts similar to transactions between the devices in bitcoins. Hence, by leveraging the use of the decentralized system, we can have peer to peer communication established without trusting the other party. So, there will be execution of transactions and messages between two companies. For example, you can envision a factory controlling the production of its items and monitoring as well without any communication to centralized system. Blockchain leverages a decentralized network of computers that conduct intermediary tasks over the Internet.

All transactions are recorded into a digital ledger, which is publicly available and fully distributed to all members of the network (so-called nodes). The distributed ledger (DL) approach eliminates the need for a trusted third party.

Figure 6.2 describes how this ledger is acting decentralized by checking automobile's safety, checking power of appliances, and registering new devices.

Hence ledgers and smart contracts are basic building blocks of blockchain technology that runs the entire system and maintains privacy in transactions as well (Khudnev, 2017).

6.5   COMPONENTS OF LEDGER

The five components of the distributed ledger are as follows:

The Network of Nodes: These nodes are the parts of the network that is used for maintaining the ledger and transactions.

Tokens: These are the units of exchange in the transactions. The basic units of networking are cryptocurrency and digital currency.

The Structure: This defines the basic framework of ledger. A blockchain consists of chained blocks that are connected in a sequence containing transaction information. The summation of all transactions forms a ledger.

The Consensus Mechanism: This mechanism basically prevents double spending of the tokens and determines the accurate value of the ledger.

Rules: These are basic rules of communication between all the parties. These rules are present on bitcoin and Ripple networks. Bitcoins are default systems for payment while Ripple is for external assets.

6.6   WORKING OF LEDGER

The blockchain as we know is a decentralized and distributed system (Kshetri, 2017) for managing the transactions and their privacy. Before discussing about blockchain, we should be aware of the structure of classic ledger. Initially, ledgers were used for maintaining and storing various transactions regarding the property in banks and government offices. In order to build integrity between two parties, banks were used as means of central authorization for acknowledging all the transactions and designing of the contractual documents between parties. Whoop the blockchain works, let us talk about the classical ledger or centralized architecture. The ledger manager had the control to view the ledgers so that parties can buy and sell. The ledgers are just visible to the manager in charge and not to the parties in communication, hence it is a fully trusted system and the third party has control over the transactions management system. Similarly, blockchains have the same functionality but they have no ledger managers. Hence there is no central authority in this system rather the companies participating have copies of the original ledger with them. The participating user can add or modify the transaction itself, but it gets added to the block only if it is verified by all the parties. An automated check of all the users is done for safe and secure transactions. Once the transactions are verified it is linked to other transactions in the queue forming a chain known as bock chain. Figure 6.2 shows the working of financial transactions in blockchain.

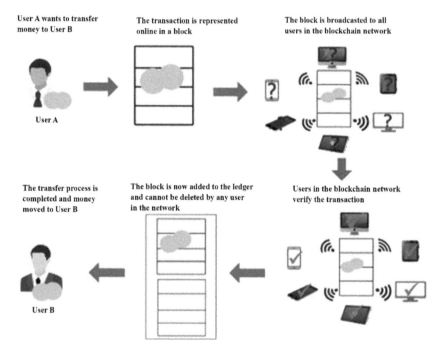

FIGURE 6.3   Financial transaction in blockchain.

In Figure 6.3, user A wants to transfer money to user B in a financial transaction.

1. Each transaction is represented in a block. After the block is created on imitation of the request by User A, it is transferred to all the participating users to verify the transaction in the block.
2. When the transaction in the block gets verified and it cannot be deleted now the transfer process will begin.
3. User B receives the money.

6.7   CHALLENGES OF BLOCKCHAIN

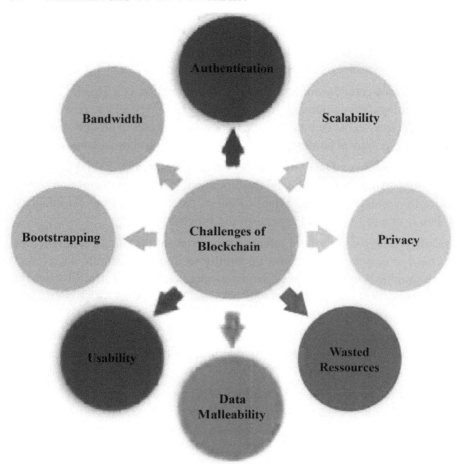

FIGURE 6.4   Applications of blockchain.

Figure 6.4 depicts the key challenges faced while creating a new ledger in blockchain and updating it.

Scalability: As we know, every transaction is stored in the distributed ledger for privacy and security. These transactions in the blocks increase daily. In order to verify the validation of the transaction, we need to maintain a ledger for examining the transaction source. The creation of a block is an important aspect of bitcoins. Bitcoins can make only seven transactions per second but we need millions of transactions in real time applications. Hence size and time play an important role in the execution of transactions. Hence an optimization technique is needed to delete older ledgers and keep new ledgers (Akhtar et al., 2016).

Privacy: The users participating in the transactions can keep their identity anonymous through private and public keys. As all the users can view the value of transactions so privacy is a challenge ahead.

Wasted Resources: Saving the energy of computation is a significant challenge in computer engineering. In blockchain technology a huge power is lost in the computation of last transactions. Hence the computing process can speed up by using graphical processing units along with central processing units.

Data Malleability: Data integrity is an important aspect of blockchain technology. An intruder attack on the block can lead to retransmission of the transaction which may lead to duplicating of data. Maintaining the integrity of data is one of the critical aspects of the blockchain.

Usability: The usability here refers to the nature of the blockchain whether it is hard or easy to use. The new upcoming technologies should have interactive graphical user interfaces for developers and clients. But blockchain consists of continuous formation of blocks and transactions which is difficult to change according to users.

Bootstrapping: In order to transfer the transactions, data and documents of blockchain need to be verified by documents. Hence this tedious process cost time and money both.

Bandwidth: The size of the block is an important aspect (Sheri, 2017) which computes the number of transactions in the block. Hence all the participants need to be informed about the size of block, the number of new transactions. A normal block has 1000 transactions having a size of 1MB.

Authentication: Authentication and security breach is one of the vital issues that affect the blockchain environment. Hence keeping the identity of user correct is the issue that is faced while authenticating bitcoins. The private keys need to be protected for each user by a secure cryptographic algorithm. In the existing blockchain, once a user identity is created, there is no guarantee that the user requesting the identity is the correct owner of that identity and not a malicious one (Atlam et al., 2018).

6.8   IOT AND APPLICATIONS

IOT can be defined as the third wave of IT industry (Alaba et al., 2017) which is a connection of intelligent devices exchanging data through a unified framework. Common IOT applications are as follows:

Smart Homes: The automation in homes is one of the popular applications of IOT where all the appliances in the home communicate for different uses. All the devices are equipped with sensors. So many malicious activities in the home are recorded by these devices which are reported to the owner of the home via message or alarm.

Health Industry: So, this is one of the major innovations of IOT devices. Devices like smartwatches can track user activity, heartbeat, and health conditions. Some of these devices can even treat the patient if it is a minor disease and in case of emergencies, it can send messages to the hospitals.

Agriculture: So IOT devices can also help in producing better crop cultivation based on weather conditions and other parameters. This helps in producing a better investment and profit for farmers. Sensors can be used to monitor and augment soil quality, rainfall, crop growth, and providing equipment help via harvesters.

Logistics: IOT devices also regulate the development of businesses and industries by using RFID and NFC technology. Hence the sensors and tags attached to devices give it real-time tracking, condition, delivery date, and transmit messages as well (Atlam et al., 2017).

Smart City: This is one of the major innovations of IOT. IOT devices, such as sensors, are used to collect information about traffic, lights, and parking. Hence it gives all information about the traffic ahead and how it can be controlled, which ways to avoid.

Fitness and Health: This is one of the most important applications of IOT devices. IOT devices, like fit bit, apple watch, tracks the health of an individual and parameters like number of hours you sleep, heart rate, number of steps taken, and other statistics related to health. This data is increasing day by day and can be used for research ahead by companies.

6.9   COMMON IOT STANDARDS AND PROTOCOLS

A common IOT framework consists of a large number of electronic devices having embedded sensors interconnected via a common bus. Figure 6.5 shows a layered framework of IOT devices (Granjal et al., 2015) and their mode of working.

In Figure 6.5, all the devices are connected via open gateways to interact with the outer world as well. This framework is used for messaging applications, routing, key authentication, and connecting physical devices. The commonly used protocols are low power wide area network (LPWAN) protocols. This is an IEEE standard 802.15.4 layer used for long-distance communications. The routing protocol (RPL) is for loss and less power networks. User datagram protocol (UDP) is for easy communication between devices. The compression is better in user datagram protocol. The constrained application protocol (CoAP) works on request-response method for less power communication. The internet control messaging protocol (ICMP) is for controlling the packets or messages that are not delivered (Granjal et al., 2015).

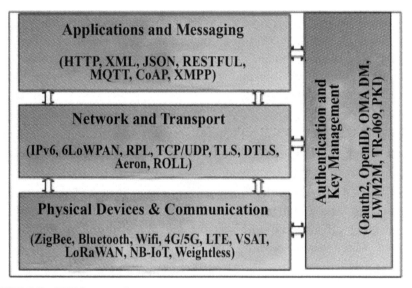

FIGURE 6.5 IOT framework.

6.10 SECURITY ISSUES IN IOT DEVICES

The security issues in the IOT devices are classified into three main classes:

Low-Level Security Issues: These issues are present at the physical and data link layer of communication.

o **Jamming adversaries:** These attacks destroy the IOT devices by emitting out radio waves. The data transmission gets affected by this attack resulting in malfunctioning in the devices.

o **Insecure initialization:** Insecure initialization of the physical layer also destroys the systems working if security is not ensured at the physical layer.

o **Low-level spoofing attacks:** The Sybil nodes in the network used MAC values which copies the IP address of the other device and affects the network too.

o **Insecure interface:** The depletion in the physical interfaces of the network can lead to a major loss in the network.

o **Sleep deprivation attack:** When a large number of tasks are programmed in IOT devices then it results in depletion of the battery.

Intermediate Level Security Issues:

These issues are at communication and session management at transport and network layers.

o **Duplication attacks due to fragmentation:** The duplication of packets results in loss of the quality of the network, restarting the devices, fragment duplication which hinders the communication between devices.

o **Insecure neighbor discovery:** All the devices neighboring the transmitted device must be discoverable so that our packet reaches the destination safely.

o **Buffer reservation attack:** This attack is a denial of service attack as the network gets congested by incomplete packets leading to exploitation by the attacker.

o **RPL routing attack:** This attack may lead to depletion of the network again due to affected nodes in the network.

o **Sinkhole and wormhole attacks:** These attacks create a tunnel inside the network via the transmission of packets leading to a denial of service and packet dropping.

o **Sybil attacks:** These attacks exist on a higher layer as well leading to fake identities in the network.

High-Level Security Issues:

These issues exist because of:

o **CoAP security with internet:** The sophisticated layer containing the application layer is also susceptible to attacks. The CoAP is a web transfer protocol for inhibited devices using DTLS bindings with various security modes to provide continuous security. The CoAP messages pursue a precise design defined in RFC-7252, which needs to be encrypted for protected communication. Similarly, the multicast holds up in CoAP requires sufficient key administration and authentication mechanisms.

o **Insecure interfaces:** The interfaces accessed through web, mobile, and other digital devices are vulnerable to various attacks affecting data security.

o **Insecure applications:** All the software needs to be tested and updated before use in the IOT devices.

6.11   CRYPTOGRAPHY IN BLOCKCHAIN

The blockchain network uses hashes in combination with digital signatures (Sicari et al., 2015) in order to ensure security in data. Whenever a transaction is launched the wallet consists of a private and a public key. The public key is generated from the private key which is basically the account number of customers. The private key is a password generated key which is a 256-bit integer. Both the keys are kept secure unless needed. The private key is always kept secure and is not shared with all participants.

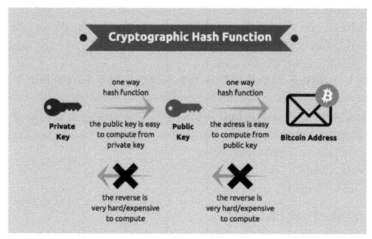

FIGURE 6.6   Relationship between keys.

There is a good mathematical relation existing between private and public keys in cryptography as shown in Figure 6.6. These relationships between public and private key are shown by cryptographic algorithms. These algorithms are single way algorithms which generate public key from a private key. Hence generating a private key from public key is difficult as it would lead large computation powers in trillions. Hence in order to generate public key from private key series of steps are involved. Hashing is the most important step in generating bitcoin address. The public key gets hashed using a secure hash algorithm. The hash string that is generated in the wallet is the final output (Granjal et al., 2008).

6.12   APPLICATIONS

Hashing: Hashing in cryptography is a mechanism that converts large unstructured data into short strings of digits. This mechanism is used in blockchain technology as well to ensure integrity while creating new transactions. It uses a secure hash algorithm (SHA-256) which converts the input into a small output string. For example, the sentence "I am studying science" looks like this 73abddd89faf78d9930e4b523ab804026310c9. If we change the sentence to "I am studying" then entire hash digest changes to 4314d903f04e90e4a505768524.

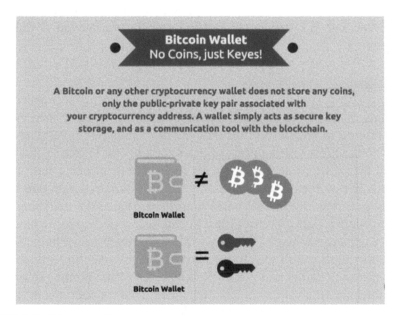

FIGURE 6.7   Bitcoin wallet.

Bitcoin Wallets: This wallet does not store any money or cash but stores the public–private key pair related to bitcoin address. A bitcoin wallet in Figure

6.7 is simply a storage area of public–private key. Whenever a user uses it bitcoin in it accesses the wallet for secure transaction. Similarly, bitcoin address is just an email address that you can access via your public key. Hence you must keep your keys and email address protected with you so that your funds are secure (Xu et al., 2016).

Digital Signatures: The bitcoin network uses digital signatures along with hashes for data privacy and security. In order to authenticate the transactions, digital signatures are used. The signature signifies a particular user is the owner of that hash. Signing both inputs and hash values are mandatory. The digital signature works as follows:

  a.   User A generated hash related to its transaction. The private key then encrypts the hash digest which signs the transaction.
  b.   This hash digest is sent to user B along with the public key generated by the private key.
  c.   User B generates another hash message and then compares it with hash generated before.
  d.   Using cryptosystem user B decrypts the fits hash to see its owner.
  e.   Then both the hashes get verified and the transaction is digitally signed.

6.12.1   MINING ALGORITHM

The mining of data from bitcoins also requires cryptosystems. The mechanism of mining in bitcoins excites the network validators to validate these bitcoins' transaction which ahead validates the bitcoin. Hence in this algorithm, all the participants compete with each other generating hash digest for a particular problem. All the hash digests are compared with one that is originally generated by the SHA-256 algorithm. Hence the matching hash user is the winner. Hence you can cheat in bitcoins by putting block before it. Figure 6.8 depicts working of the mining algorithm in bitcoins. In the figure, it shows how blocks are added to the network one by one. So, validation of the hash message by the miner is an essential step of the mining algorithm. Hence miners are the validators that authenticate the transactions in mining algorithms. The SHA-256 algorithm uses hash messages for each participant individually that miners validate. After a thorough comparison is done, the message that matches the original hash digest is chosen for validation. This algorithm is an essential milestone that will be used for implementing blockchain algorithms ahead (Atlam et al., 2017).

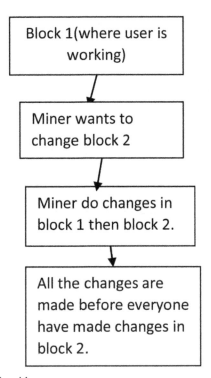

FIGURE 6.8   Mining algorithm.

Figure 6.8 describes the mechanism of mining algorithm which has four important steps:

1. Selecting the block in the blockchain where the user wants to work.
2. Then the user is validated as a miner who validates the transaction.
3. Miner has the authority to change the block.
4. So, all the users would change in a new block first as validated by the miner (Lamport et al., 1982).

6.13   CHALLENGES OF SECURITY IN IOT AND BLOCKCHAIN

While ensuring security and integrity in IOT devices through blockchain many challenges are being faced:

Big Data: Big data as we know is a date made up of large volumes of meta-data and structured data as well. Extracting and mining information from this data is quite a challenging task. The preprocessing and cleaning of this data is quite a challenging task as it consists of a huge volume of data. The IOT devices also generate millions of data which needs to be handled accurately for security and integrity issues. Hence the security of data is an important aspect that needs to be considered.

Network: The key aspect of the IOT devices is the framework which connects them. As we know these devices are of different functionalities using different sensors and protocols. Hence choosing the accurate topology and protocol is an important decision to be made before ensuring security in devices.

Feature Selection: The IOT and blockchain is one of the discussed issues which have a large number of devices connected via a common framework. In order to extract different features from these devices, we have to consider communication patterns, network quality, and other parameters for achieving high latency and efficient communication.

Interoperability: This is the biggest challenge in the IOT devices. As we know different IOT devices have different specifications, communication patterns, and network protocols. Hence all the devices in the IOT network must be compatible with each other. For this purpose, a separate network layer has also been designed for such devices.

Scalability: As we know the number of devices in the blockchain and IOT network are increasing and so is the data within them. The scalability of the system should be handled effectively as the number of devices in the system increases.

Security and Privacy: As the number of devices in IOT network is increasing so is their security. In order to achieve security in these devices' user data needs to be protected from malicious users.

6.14   BLOCKCHAIN MODEL FOR CRYPTOGRAPHY

Hawk is a decentralized system for cryptography (Wood, 2017) replacing bitcoins for trusted financial transactions by using efficient cryptographic algorithms. Hawk builds a private network in blockchain which can be written by any nonprogrammer as well. It acts as an interface between user and blockchain technology. Figure 6.9 explained working of hawk (Lamport et al., 1982).

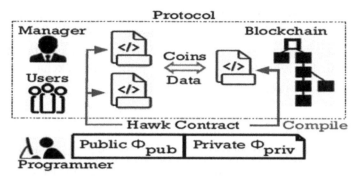

FIGURE 6.9 Working of hawk.

A hawk is divided into two parts as shown in Figure 6.9:

1. **Private Part:** The private section is denoted by $\emptyset_{priv}$ which takes in input data taken by the system, that is, data and currency and the output data we get after computation by cryptographic algorithms. Hence private hub protects the data and currency from external users and currency as well.

2. **Public Part:** The Public section is denoted by $\emptyset_{pub}$ and does not modify or read private data and money.

6.15 CRYPTOGRAPHIC SPECIFICATIONS IN HAWK

The blockchain is a decentralized system which is a group of miners which is deploying cryptographic algorithms to maintain security and privacy. Blockchain hence is an abstraction of reliability, privacy, and trust which maintains a ledger storing transactions and user-defined data (Liu et al., 2015). The important components of the model are:

Time: The time is denoted by epochs and rounds. The clock is discrete in nature.

Public state: This means all the users can access the data. This means a user can access the public ledger and transactions in the blockchain program.

Delivery: The delivery of the message follows a round-robin pattern where the message arrives at the next round and reorder messages. Therefore, a reliable channel is followed having sufficient redundancy

Pseudonyms: These are the polynomial variables generated when users react with blockchain network.

Reliability: The above assumptions tell that blockchain is a reliable network for secure computation.

Programmers, wrappers, and functionalities: These are the ideal functionalities defined in the blockchain program. These are the programs written in pseudonyms having an ideal program, blockchain program and User program. The following wrappers are defined in blockchain:

Ideal wrapper F: This function transforms the Ideal P program into UC ideal functionality.

Blockchain wrapper G: This function transforms blockchain program (B) to blockchain functionality to G and then models the program accordingly.

Protocol wrapper Π (P): This transforms a user program (P) into user side Π (P) using a user manager protocol.

Conventions for writing programs: The wrapper-based functions use user-defined programs, blockchain programs, and user protocols. The following components are used:

Timer activation points: The timer activation points consist of functionality wrapper and blockchain wrapper. The clock in the blockchain invokes the timer activation point.

Delayed processing: When writing the blockchain program, the program gets delayed due to wrapper functions. All the programs are written in gray code which means it takes place after the first round. Hence ideal cash programs used for processing of messages using the timer operation.

Pseudonymity: The part identifiers are stored in ideal programs, blockchain programs, and user side protocols. So when we type the message "upon receiving message from *some P*", this means it accepts message from P. when we type message "send *m* to $G(B)$ as nym *P*" inside a user program, it means this sends an internal message ("send", *m*, *P*) to the protocol wrapper function. The protocol wrapper function will authorize the messages appropriately (Kshetri 2017).

The Hawk program below is used for implementing a sealed, second-price auction where the highest bidder wins but pays the second highest price. Second price auctions basically do truthful bidding under certain circumstances and it is important that bidders submit their bid values privately under private part. The Hawk program consists of a private portion φ_{priv} that computes the winning bidder and the price to be paid and a public portion φ_{pub} that will rely on public deposits to defend bidders from a manager. The assumption is made that all the bidders are apriori. The program for blockchain model works as follows:

```
HawkDeclareParties (Seller, /* N parties */);
HawkDeclareTimeouts (/* hardcoded timeouts */);
// Private portion φpriv
private contract auction (Inp &in, Outp &out) {
int winner = −1;
int bestprice = −1;
int secondprice = −1;
for (int i = 0; i < N; i++) {
if (in.party(i).$val > bestprice) {
secondprice = bestprice;
bestprice = in.party(i).$val;
winner = i;
else if (in.party(i).$val > secondprice) {
secondprice = in.party(i).$val;
}
}
// Winner pays secondprice to seller
// Everyone else is refunded
out.Seller.$val = secondprice;
out.party(winner).$val = bestprice-secondprice;
out.winner = winner;
for (int i = 0; i < N; i++) {
if (i != winner)
out.party(i).$val = in.party(i).$val;
}
}
// Public portion φpub
public contract deposit {
// Manager deposited $N earlier
def check(): // invoked on contract completion
send $N to Manager // refund manager
def managerTimeOut():
for (i in range($N)):
send $1 to party(i)
}
```

6.16 GB FUNCTIONALITY BLOCKCHAIN IMPLEMENTATION

Given: A blockchain program denoted B, the $G(B)$ functionality is defined
as below:

Init: Upon initialization, perform the following:

A ledger data structure ledger (^-P) stores the account balance of party ^-P. Send the entire balance ledger to A.

Set current time $T := 0$. Set the receive queue rqueue $:= \varnothing$.

Run the Init procedure of the B program.

Send the B program's internal state to the adversary A.

Tick: Upon receiving tick from an honest party, if the functionality has collected tick confirmations from all honest parties since the last clock tick, then apply the adversarial permutation perm to the rqueue to reorder the messages received in the previous round.

Call the timer procedure of the B program.

Pass the reordered messages to the B program to be processed. Set rqueue $:= \varnothing$.

Set $T := T + 1$

Other activations:

- Authenticated receive: Upon receiving a message (authenticated, m) from party P:

Send (m, P) to the adversary A

Queue the message by calling rqueue.add(m, P).

- Pseudonymous receive: Upon receiving a message of the form (pseudonymous, m, $^-P, \sigma$) from any party:

Send $(m, {}^-P, \sigma)$ to the adversary A

Parse $\sigma := (\text{nonce}, \sigma\_)$, and assert verify(^-P.spk, (nonce, T, ^-P.epk, m), $\sigma\_) = 1$

If message (pseudonymous, m, $^-P, \sigma$) has not been received earlier in this round, queue the message by calling

rqueue.add(m, ^-P).

- Anonymous receive: Upon receiving a message (anonymous, m) from party P:

Send m to the adversary A

If m has not been seen before in this round, queue the message by calling rqueue.add(m).

Permute: Upon receiving (permute, perm) from the adversary A, record perm.

Expose: On receiving expose state from a party P, return the functionality's internal state to the party P. Note that this

also implies that a party can query the functionality for the current time T.

Ledger operations: // *inner activation*

Transfer: Upon recipient of (transfer, amount, ^-Pr) from some pseudonym ^-Ps:

Assert ledger(^-Ps)$\geq$ amount

ledger(^-Ps) := ledger(^-Ps)$-$ amount

ledger(^-Pr) := ledger(^-Pr) + amount

6.17 POTENTIAL BLOCKCHAIN SOLUTIONS FOR IOT SECURITY

In order to make IOT devices more secure smart contracts are being deployed in order to monitor security. Some of the solutions adopted to monitor security in IOT devices are:

Smart Contracts: A Smart Contract is a computerized contract having its own protocols and rules (Mitchell and Chen, 2014). These are programs written in JavaScript language and uploaded on blockchain. Ethereum in smart contracts provides security through cryptographic algorithms like the RSA algorithm and SHA_256. An Ethereum consists of ether as its digital currency for augmenting security in blocks. Figure 6.10 shows the framework of Ethereum in a blockchain.

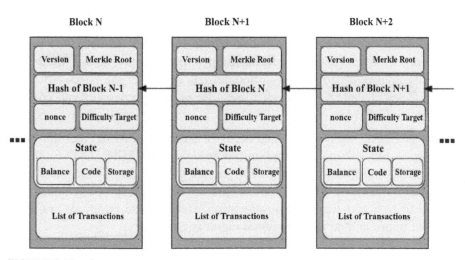

FIGURE 6.10 Smart contracts.

Figure 6.10 shows the smart contract and its framework. A smart contract has its own address, structure, and ether coins.

Space: The address space used in blockchain must be 160 bits. A blockchain uses 160-bit hash which can hold up to $1.46*10^{48}$ devices. Hence with the use of blockchain we do not require a centralized system rather a decentralized system allocates global unique identifier (GUI) to IOT devices in the network. The address space has more storage space than IPV6 and is more scalable.

Governance: The identity of an IOT device is the most challenging issue to be dealt with blockchains. Hence the properties of a device include its model no, serial no, date of manufacturing, software version, owner name, GPS coordinates, IMEI no. As the owner of the device changes, these properties may vary ahead. Blockchain can solve this issue as well by updating its information online in the ledger using a decentralized system. The trust chains are being used by blockchains for holding the information correctly.

Data Integrity: Data that is present in the IOT devices need to be kept secure by using cryptographic algorithms that encrypt the data using the public key. The data that is encrypted is updated on the ledger. Hence all the transactions are made secure by ensuring proper encryption.

Authorization of Data: Blockchains can provide more secure data transmission and storage using smart contracts, public–private key pair, hashing, and mining algorithms. The traditional algorithms being used like Role Based Access Management (RBAC), OAuth 2.0, OpenID, OMA-DM, and LWM2M are centralized in nature and more complex. Hence, we prefer a decentralized mechanism for communication and transmission. Smart contracts give rights to people to update data and reset the passwords for the devices and some rights to other users for public–private key encryption (Roman et al., 2014).

Security: Hence, we conclude that security is more in IOT devices through the use of blockchains by applying protocols like HTTP, COAP, TLS, 6LOWPAN, and PKI. These protocols ensure safe routing of the messages between IOT devices. With PKI public–private key is also properly managed leading to safe and secure communication. Hence these protocols are less complex to use and do not take much memory as well.

6.18   OPEN CHALLENGES AND FUTURE RESEARCH DIRECTIONS

Despite deploying robust approaches for securing our devices in the network by leveraging blockchain cryptography techniques and other network security protocols still security is an open issue in IOT devices and hindrance as well. The challenges that still exist are:

Resource Constraints: This is the main hindrance that exists while transmitting data and exchanging keys while broadcasting messages to other participants in the network. Hence all the algorithms and protocols used must be less complex and acquire less computation power while transmission of data.

Heterogeneity: An IOT framework consists of low power devices to high-power devices in the network. So, the blockchain framework should be flexible and dynamic to adapt to all type of devices deployed in the network.

Interoperability: This is again an important issue that needs to be considered while deploying security in IOT devices. Hence all the devices and protocols must be interoperable at all levels (Cirani et al., 2014).

Single Point of Failure: Maintaining decentralized architecture in IOT devices is again a hindrance. As it consists of large number of devices ranging from low power to high power maintaining the decentralized architecture is necessary to avoid a single point of failure.

Hardware Failure:

With the rise of low power internet devices, hardware failure, and failure in network security protocols need to be verified. Hence all the devices must be tested before deployment in network for any vulnerability.

Other Issues: The other issues that exist while deploying security in IOT devices are trusted updates of the devices. With blockchain this can be verified by verifying the governance of updates (Nakamoto, 2009).

6.19   CONCLUSIONS

Recently we could see that blockchain technology has emerged worldwide and received extensive interest in terms of security. It has the potential to solve many problems in all types of industries. Still, it is in the initial stages of development many challenges still exist as we have discussed in this chapter. Blockchain has involved in almost every industry even if it still in the first stage of approval to solve the problem of centralization, security, network communication, collecting data, cryptography, and hashing the data. All of this data is maintained in a ledger in a block, hence for integration of IOT with blockchain, this chapter presented presents a discussion of the technical aspects of security in blockchain and IoT. All the protocols related to IOT were discussed and how security can be breached. In order to make IOT devices, secure deployment of blockchain and its architecture was discussed briefly. But still, we could see security is a hindrance in IOT devices due to immature practices of software deployment. Hence for proper security in the devices, blockchain should be well distributed at all the layers of the framework.

KEYWORDS

- **Rivest–Shamir–Adleman**
- **blockchain**
- **cryptography**
- **internet of things**
- **cryptosystem**

REFERENCES

Abduvaliyev, A.-S. K.; Pathan, J.; Zhou, R.; Roman, W.-C.; Wong. On the Vital Areas of Intrusion Detection Systems in Wireless Sensor Networks. *IEEE Commun. Surv. Tutor.* **2013**, *15* (3), 1223–1237. http://dx.doi.org/10.1109/SURV. 2012.121912.00006.

Akhtar, F.; Rehmani, M. H.; Reisslein, M. White Space: Definitional Perspectives and their Role in Exploiting Spectrum Opportunities. *Telecommun. Policy* **2016**, *40* (4), 319–331. http://dx.doi.org/10.1016/j.telpol.2016.01.003.

Alaba, F. A.; Othman, M.; Hashem, I. A. T.; Alotaibi, F. Internet of Things Security: A Survey. *J. Netw. Comput. Appl.* **2017**, *88* (Suppl. C), 10–28. http://dx.doi.org/ 10.1016/j.jnca.2017.04.002.

Atlam, H. F.; Alenezi, A.; Alassafi, M. O.; Wills, G. B. Blockchain with Internet of Things: Benefits, Challenges, and Future Directions. *Int. J. Intell. Syst. Appl.* **2018**, *41* (10), 40–48.

Atlam, H. F.; Alenezi, A.; Walters, R. J.; Wills, G. B. An Overview of Risk Estimation Techniques in Risk-based Access Control for the Internet of Things. In Proceedings of the 2nd International Conference on Internet of Things, 2017, Big Data and Security (IoTBDS), 2017, pp. 254–260.

Butun, I.; Morgera, S. D.; Sankar, R. A Survey of Intrusion Detection Systems in Wireless Sensor Networks. *IEEE Commun. Surv. Tutor.* **2014**, *16* (1), 266–282. http://dx.doi. org/10.1109/SURV.2013.050113.00191.

Cirani, S.; Ferrari, G.; Veltri, L. Enforcing Security Mechanisms in the IP-based Internet of Things: An Algorithmic Overview. *Algorithms* **2013**, *6* (2), 197–226. http://dx.doi. org/10.3390/a6020197.

Granjal, J.; Monteiro, E.; Silva, J. S. Security for the Internet of Things: A Survey of Existing Protocols and Open Research Issues. *IEEE Commun. Surv. Tutor.* **2015**, *17* (3), 1294–1312. http://dx.doi.org/10.1109/COMST.2015.2388550.

Granjal, J.; Silva, R.; Monteiro, E.; Silva, J. S.; Boavida, F. Why is IPSec a Viable Option for Wireless Sensor Networks. In 2008 5th IEEE International Conference on Mobile Ad Hoc and Sensor Systems, 2008, pp. 802–807. http: //dx.doi.org/10.1109/MAHSS.2008.4660130.

Khudnev, E. Blockchain: Foundation Technology to Change the World. *Int. J. Intell. Syst. Appl.* **2017**.

Lamport, L.; Shostak, R.; Pease, M. The Byzantine General's Problem. *ACM Trans. Program. Lang. Syst.* **1982**, *4* (3), 382–401.

Liu, C.; Wang, X. S.; Nayak, K.; Huang, Y.; Shi, E. Oblivm: A Programming Framework for Secure Computation. In *2015 IEEE Symposium on Security and Privacy* 2015, pp. 359–376).

Kshetri, N. Blockchain's Roles in Strengthening Cybersecurity and Protecting Privacy.. *Telecomm. Policy* **2017**, *41* (10), 1027–1038.

Mitchell, R.; Chen, I.-R. Review: A Survey of Intrusion Detection in Wireless Network Applications. *Comput. Commun.* **2014**, *42*, 1–23. http://dx.doi.org/ 10.1016/j. comcom.2014.01.012.

Nakamoto, S. Bitcoin: A Peer-to-Peer Electronic Cash System, Available: http://www. bitcoin. org/bitcoin.pdf, 2009.

Roman, R.; Alcaraz, C.; Lopez, J.; Sklavos, N. Key Management Systems for Sensor Networks in the Context of the Internet of Things. *Comput. Electr. Eng.* **2011**, *37* (2), 147–159. Modern Trends in Applied Security: Architectures, Implementations and Applications.

Stanciu, A. Blockchain Based Distributed Control System for Edge Computing. In 21st International Conference on Control Systems and Computer Science Blockchain, 2017, pp. 667–671.

Sicari, S.; Rizzardi, A.; Grieco, L.; Coen-Porisini, A. Security, Privacy and Trust in Internet of Things: The Road Ahead. *Comput. Netw.* **2015**, *76* (Suppl. C), 146–164. http://dx.doi. org/10.1016/j.comnet.2014.11.008.

Wang, Y.; Uehara, T.; Sasaki, R. Fog Computing: Issues and Challenges in Security and Forensics. In 2015 IEEE 39th Annual Computer Software and Applications Conference; Vol. 3, 2015, pp. 53–59. http://dx.doi.org/10.1109/ COMPSAC.2015.173.

Wood, G. Ethereum: A Secure Decentralized Transaction Ledger (2014), 2017.

Xu, K.; Qu, Y.; Yang, K. A Tutorial on the Internet of Things: From a Heterogeneous Network Integration Perspective. *IEEE Netw.* **2016**, *30* (2), 102–108.

Yi, S.; Qin, Z.; Li, Q. Security and Privacy Issues of Fog Computing: A Survey. In Wireless Algorithms, Systems, and Applications the 10th International Conference on, 2015, pp. 1–10.

CHAPTER 7

Blockchain Technology and Internet of Things

SUMATHY ESWARAN* and S. P. RAJAGOPALAN

Dr. MGR Educational and Research Institute, Chennai, Tamil Nadu, India

Corresponding author. E-mail: sumathyeswaran@drmgrdu.ac.in

ABSTRACT

Internet of Things (IoT) devices are projected to exceed $20 billion by 2020. Blockchain technology (BCT) business is estimated to be at $20 billion by 2024. As of now, 45% of industries are adopting IoT, 69% of banks are experimenting with BCT, and 33% of banks are expected to adopt BCT by next year. Reported savings to banks on adoption of BCT are likely to be around $10 billion. Hence, the authors believe it is time to discuss these two trending technologies.

This chapter provides valuable information and equips one with an understanding on how the two trending technologies, BCT and Internet of Things (IoT), revolutionize transactions and business networks. The topics are handled in a chronological manner so that the reader gets a clear picture after reading.

The chapter starts with blockchain (BC) fundamentals, how BC works, and its applications. Then, the noncontroversial, most popular application of BC, Smart Contracts and their place in industry are highlighted.

Next, IoT and its ecosystem are introduced with the road map of IoT; applications of IoT are dealt along with the issues in IoT environment. How IoT brings in smartlife and changes the way the world does business are discussed. Industrial IoT (IIoT) is a catalyst for Industry 4.0 enabling smart manufacturing. The biggest concern in IoT ecosystem is security and privacy. IoT devices are susceptible to Distributed Denial of Service and linking attack. These attacks and suggested solutions are detailed.

Synergy is brought in when solutions are given as IoT and BC combined. Smart assets are transacted in a decentralized system instantaneously in a verifiable, untamperable manner. The trust and authority are moved away from centralized agency to the technology in the hands of the participants. In fact, few experts believe, that these two technologies will become inseparable. IBM believes device independence is achieved with BCT–IoT combined. There are other technologies, such as Cloud, Big Data, fog computing, 5G, IOTA, and so on, which give a big fillip to IoT and bring IoT closer to life.

This chapter is a good step to highlight, how to achieve a smooth, secured smartlife with IoT–BCT combined and the state of current research in this direction.

"What the Internet did for Communications, Blockchain will do for Trusted Transactions."

—Ginni Rometty

"The Internet of Things has the potential to change the World, just as the Internet did. May be even more so"

—Kevin Ashton

7.1   BLOCKCHAIN TECHNOLOGY

The global spending on blockchain (BC) solutions in 2018 is $2.1 billion (Raja, 2018). Expected growth of BC market by 2024 is worth $20 billion. It is important to note that 69% of banks are currently experimenting with blockchain technology (BCT) and 33% of bankers are expecting commercial BC adoption by 2019. When banks adopt BCT, the potential savings for banks is around $8–$12 billion.

7.2   BLOCKCHAIN FOR BUSINESS

McKinsey, in its report (Carson et al., 2018) on BC for business value, defines BC as a distributed shared ledger shared amongst the participating nodes over the Internet, with addition of encrypted information to the chain-holding historical records, after consensus and validation by the members.

7.3   WHAT IS BLOCKCHAIN?

BC-shared ledger can be used for recording and tracking any transaction of tangible or intangible assets with values (wealth). BC is not a replacement for traditional database but contains verified proof of transactional information. BC can be used in any industry where transactions of assets happen. The innovation of BC technology is triggered by the advancements in software engineering, distributed computing, cryptography, and game theory (Sultan et al., 2018).

BC uses distributed ledger technology (DLT) that makes enterprise practices more efficient by recording and tracking assets and benefiting with increased levels of trust, accountability, immutability, and transparency. DLT is a decentralized shared ledger synchronously kept up at participating nodes on the Internet, but entries, that is, blocks are added/recorded with consensus of the nodes, doing away with the control of the third-party organization. The shared ledger records are verifiable by the authorized members, secure because encrypted before recording, transparent with timestamp, and other details. Reversal of a transaction can be done only by another new transaction. Thus, BC for business becomes immutable ledger for recording transactions.

7.4   BLOCKCHAIN FORMATION

BC, as the name implies, stores transaction data in blocks; every block is chained to the earlier block in the chain except the first one, in the order of timestamp, to form a chain. As the number of transactions increases, the length of the chain also grows unidirectionally as in Figure 7.1.

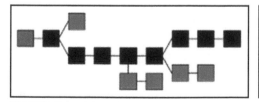

	Green – Initial block or Genesis Block
	Black – New blocks linked by time to Blockchain
	Purple –Orphan blocks existing outside main chain

FIGURE 7.1   Blockchain formation.
Source: Wikipedia CC BY 3.0 hide terms, file: Blockchain.svg, created: January 16, 2011.

7.4.1 BLOCK GENERATION

A block consists of four components; those are Transaction Root, Previous Block Hash, Timestamp, and the Nonce. A block can have a number of related transactions. Each transaction has its own hash. Transaction root is also called Merkle Root, which is the mother of all the hashes in the block. Every block generates a hash. The hash from the previous block is used in generation of current block, ensuring immutability. Nonce is a number used by miners to show proof of work in solving the puzzle in bitcoin generation. However, in the case of Smart Contracts, Nonce has less relevance and can be used as a token number for communication among members.

7.5 BLOCKCHAIN FUNCTIONAL CHARACTERISTICS

The functional characteristics of BC are immutability, decentralized, and consensus-driven auditability and transparency. These characteristics facilitate a secure and trusted transaction chain for the custody of assets (Sultan et al., 2018; Zheng et al., 2018).

Immutable: Each transaction will have to be confirmed and recorded by the participating nodes. No block in the chain can be distorted or misrepresented. Because of this immutable property, tampering is less likely. If the event tampering happens, it is traceable. Thus, BC transactions are secure and indelible.

Decentralized: Decentralized distributed network operates on P2P model of authentication, eliminating the need of central agency for authentication and the centralized server. Also, the bottleneck and single-point failure risk on centralized server are avoided. Further, distributed environment improves sustainability for the same reasons.

Consensus driven: This is the trust creation mechanism. The BC ecosystem has an identified consensus model and this consensus protocol which will be followed by each of the members while validating the block independently. This consensus model helps in replacing the central agency and moves the trust component of the transaction to the members.

Anonymity: As there is no central party, to a certain extent, anonymity of users is possible. This depends on the kind of BC network.

Transparency and Auditability: Not only each of the transactions in BC is validated and recorded with timestamp but also full transaction history can be accessed by members anytime. This is the system of transparency and auditability.

7.6   TAXONOMY OF BLOCKCHAIN

Any transaction ledger maintenance is about the members, permission for read/write, and maintaining the ledger. This has been defined by Birch and quoted in Neocapita (2017) as represented in Figure 7.2.

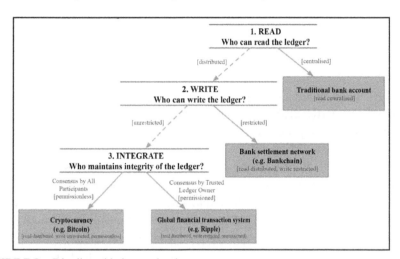

FIGURE 7.2   Distributed ledger technology.

Source: Based on Birch (2016) cited in "Distributed ledger technology: beyond blockchain," Government Office for Science, Government of the United Kingdom.

Alternatively, BryanFord (2016) states that the essence of BC is about the way the three issues, that is, a distributed ledger, consensus protocol, and membership protocol are designed in the system. Consensus protocol is about the nodes participating in validating the block and the procedure followed for validation. This follows the principle of Byzantine Fault Tolerance mechanism. Zheng et al. (2018) summarize the general methods for consensus as *Proof of Work (PoW)*, *Proof of Stake (PoS)*, *Delegated PoS*, and *Ripple and Tender mint*. Based on these characteristics of the BC, the broad taxonomy can be put into three classifications, namely, *public or permissionless*, *private or permissioned*, and *consortium or hybrid*.

Sultan et al. (2018) provide the definition of these three categories of BC as below:

Public BC (also called *permission less*) is visible to all, all can participate in the consensus process, fully distributed and has no sole ownership, anyone can create a block based on *PoW*, consensus process is open to all and everyone is eligible to participate, and public BC is fully decentralized. Crypto-currencies are an example of this category. The characteristics of public BC can be summarized as slower, open, transparent, trust-free, and illegal.

Private BC (also called *permissioned*) uses privilege settings for read/ write operations on BC because it is privilege based, mining is not required; however, consensus is optional; has single ownership to control block creation; private BC is said to be faster, managed, trusted, and legal.

Hybrid BC (also called *Consortium)* is public but to a member group, partly centralized and partly distributed, and consensus process happens on privileged servers applying the set of rules agreed by consortium. BC is shared only among entitled participants.

Zheng et al. (2018) project the comparison of these three taxonomies as in Table 7.1. Different taxonomies of BC suit different needs and objectives. The chosen design of "way of blockchaining" is dependent on the number of organizational and strategic parameters (Buterin, 2015). The division between *Permissioned* and *Permissionless* is about the rights to the validators of the block (Swanson, 2015). It is not about which is a better approach, rather it is about solving different problems.

TABLE 7.1 Comparison of Characteristics of BC Categories.

Characteristic	Public BC	Private BC	Consortium BC
Consensus determined by	All miners All members	Prime organization	Consortium members
Read permission for blocks	All members	Eligible members are identified while designing the architecture	
Immutability	Nearly impossible to tamper; highest level of immutability	Aka centralized; hence, immutability is subjective	Partly centralized; hence, immutability is subjective
Resource requirement	Extremely high	Reasonable	
Centralized architecture	Decentralized	Centralized	Partially centralized and partially decentralized
Consensus process	Permission less	Permissioned	

7.7   BCT APPLICATIONS

The decentralized transparent and secured characteristics of transactional system enabled by BCT removed few business frictions and paved way for many business applications like Global Payments, Supply Chain, Digital Assets Rights, Corporate Governance, Healthcare, Social Institutions, Democratic Participation like Voting Systems and capital markets, and so on (Wright and De Filippi, 2015). Broadly, these applications can be categorized as cryptocurrencies, commercial applications, and IoT-based Smart Life.

7.7.1   CRYPTOCURRENCIES

Cryptocurrencies are digital assets, used as a digital currency for payment during exchange of assets. It is a nongoverning currency. Bitcoin is the first use case of BCT. The legitimate owner of the coin is identifiable with precision at any time and the continuous records of currency are of importance for these cryptocurrencies to be recognized. Bitcoin-like cryptocurrencies have found their place in global currency market, with consensus-based *Proof of Work* mechanism. These currencies have replaced Centralized Server in money transaction (Back et al., 2014). Ethereum, Monero, Litecoin, NXT, ZCASH, Ripple, Dogercoin, and so on are few other cryptocurrencies amongst many.

7.7.2   COMMERCIAL APPLICATIONS

Industry uses BCT for two purposes. As the cryptocurrencies are largely getting recognized, industry uses it for in lieu of money transfer. Second best is the use of BCT as an IT enabler of transparent, immutable and traceable solutions for day to day functioning. Table 7.2 gives a few examples of use cases and the role of BCT.

TABLE 7.2   Use Cases of BCT.

Application	Applicable on	Benefit of using BCT
Smart Contracts	Any digital asset of value	A market place for transparent, reliable, third-party free transactions
Voting Systems	Citizenship identification	Avoiding manipulation of votes in democracy

TABLE 7.2 *(Continued)*

Application	Applicable on	Benefit of using BCT
Digital Identity	Driving licenses, passport, voter identity, citizenship cards	Reducing counterfeit fake identities
IoT	Smart vehicle, smart city, smart traffic	Smart data into transactional applications
Banking	Currencies, loans transactions, monetary tools, multiparty transactions	Secure and safe transactions
Manufacturing	The chronology of a product and parts that go into making the product like automobiles with their procurement and batch details	Health of the product and service
Supply chain	Stocks management, truck movement	Logistics tracking, etc.
Healthcare	Patient health records, treatment records	Automation with improved confidentiality
Publishing	Patents, copyrights, digital rights, trademarks, intellectual property rights, research publications, etc.	Ownership authentication and identification
App development	Version management and uploads in play stores	Valid ownership tracking and holding
Shares and metals trading	Ownership change	Authentication and secured trading history
Private documents	Personal certificates, property documents, and notary documents	Valid proof of ownership

7.8 BOT CHALLENGES

In spite of the benefits like immutability, decentralized, transparency, consensus driven, optional anonymity, auditability, and so on, there are few technical challenges to BCT. These are high-power consumption by hardware, powerful configuration of hardware required, data replication in distributed nodes increases the space requirement as compared to database solutions in centralized environment, transactions, or mining is slow due to distributed nodes participation (Gatteschi et al., 2018).

BCT is yet to be widely accepted by regulators (Carson et al., 2018). This is a concern for commercial applications to have easy go with BCT. In addition, there are other issues like (a) Smart Contracts cannot rely on external application programming interfaces (APIs) limiting the application development pace and bugs management. (b) Migration costs to BCT are costing the pocket. (c) Any asset has to be digitized or should be IoT enabled before bringing into the purview of BC Smart Transactions. However, IoT devices are vulnerable to attack although BC is immutable. (d) The regulators are concerned with the Darkweb activity facilitated by BCT and also the illegitimate transactions which evade state regulators.

7.8.1   SMART CONTRACT

Traditionally, most of the businesses on the exchange of assets happen over some kind of centralized ledger based on Database Management Systems. These centralized ledgers are maintained by one or more agencies involved. Some of the example applications that traditionally use centralized ledger are banking system, property transfer, vehicle sales, insurance, and so on. Robert Sams, CEO, Clearmatics, while discussing about why not centralized ledger technology, reason the three ills of the system as Sin of Commission, Sin of Omission, and Sin of Deletion. The results of these ills are Forgery of Transaction, Censorship of Transaction, and reversal/deletion of Transaction, respectively. The Smart Contract based on BCT resolves these three sins. Forgery of transaction can be prevented by cryptography and secure key management of the BCT architecture. However, Sin of Omission and Sin of Deletion are mutually exclusive in the shared ledger concept of BCT architecture and prioritized based on the use case (Sams, 2015). The finality property of distributed ledger ensures transactions are immutable and that any correction or undo will have to be done by another valid transaction recorded in the chain. The validation by legal members of the BC in DLT ensures there is no single ownership and hence no censorship.

7.8.2   SMART CONTRACT DESCRIPTION

Smart Contracts create self-enforcing contracts between two or more participants as per the set rules of the transactions, independent of the judicial and executive system of the domicile. Consensus, Provenance, immutability, and finality characteristics of BCT are inherited by Smart Contracts.

Tim Swanson in his book on great chain of Numbers defines Smart Contracts as Computer Protocols representing commercial agreement and is a program code that gets executed to facilitate and validate a transaction (Swanson, 2014). Richard Gendal Brown views Smart Contract as an event-driven program with states. This program runs on a replicated, shared ledger. The purpose is to take custody of assets listed in the shared ledger (Brown, 2015). The program resides in the replicated shared ledger and makes the rules same for everyone who uses the ledger. It is this program that incorporates the rules of the transaction and cryptographic hash values.

The sequences of steps that go into the execution of Smart Contract are as follows:

(a) **Triggering the transaction:** A wants to sell a car and setting the rules of sale.
(b) **Validation and verification:** As per the rules of business, the transaction happens between buyer and seller; verified and approved by the rules of the bank, insurance or so. Also, consensus is part of this phase.
(c) **Creating a new block:** Above verified and validated record, along with the hash of the previous block, new block is created.
(d) **Adding the block to the chain:** The created new block is communicated to all participating nodes to be appended to the shared ledger which is appended.

7.9   SUITABILITY OF BLOCKCHAIN FOR BUSINESS

Although BC for cryptocurrencies is an anonymous permissionless network, BC for business is mostly private, permissioned model, or a consortium (hybrid) model. The essential functional components of BC for a business application are (a) Shared Ledger, which is a Permissioned append-only distributed system of record, replicated and shared across the business network; (b) permissions to be set for the participants with regard to visibility of records, authentication, verification rights, and so on; (c) Smart Contract, which has the business rules built-in and executed automatically as part of transactions including cryptography; and (d) consensus mechanism required for committing transactions which is a stamp of verification and finality of transaction by members. There are many consensus mechanisms available. Chosen one is implemented (Gupta, 2018).

7.10 DIFFERENCE BETWEEN TRADITIONAL CONTRACT AND SMART CONTRACT TRANSACTION

Assets of value are transferred or exchanged in day-to-day life amongst its owners. Traditionally, assets transfer involves minimum three parties, that is, the seller/owner, buyer/beneficiary, centralized authority/authorities who approve this. In some cases, more than one authority could be involved depending on the applicable regulations for the transaction.

This can be better explained with vehicle ownership transfer. Generally, the manufacturer ties up with dealer for selling the goods. The dealer has many options to sell. One such is selling through leasing to make the purchasers purse affordable. The lessee approaches the leasing company for various options available for him. The terms are dealt and sale is done. Further, if the vehicle life is expired as per the manufacturer, it is sent to scrap merchant. In all these, the parties involved maintain their own ledger for stock, stock transfer, prospects list, vehicle information, and so on. For sale of a vehicle, licensing authority, insurance authority, and banks may get involved as Regulator. Licensing authority may also monitor the condition of the vehicle that it is fit for road conditions. These details get passed on on-demand or on an event trigger from one another causing delay and bureaucracy. Figure 7.3a represents this centralized traditional transaction.

The scenario with Smart Contract is really smart. As seen the parties involved are called nodes connected on the Internet. These stakeholders share the ledger with required entries amongst them. The Smart Contract has the rules of the transaction and encrypted and embedded. The rules of information access amongst the stakeholder are set. Also, the centralized authority is eliminated, because eligible party can see the desired information. The consensus amongst members validates the transaction. Figure 7.3b is a representative diagram for BC-enabled transaction. All these can happen on the network and almost instantaneously. This system provides transparency, authority, validity, auditability, and finality increasing the trust amongst the members.

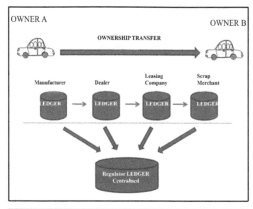

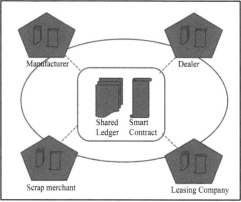

FIGURE 7.3   Transaction in car life cycle. (a) Centralized system and (b) distributed ledger using blockchain.

7.11   ADVANTAGES OF BLOCKCHAIN FOR SMART CONTRACTS

Trust is basis of business transactions because assets are exchanged. However, BC distributes the ownership of trust amongst the members in the chain. Thus, you do not need to trust the one with whom you do transaction as it is enforced in the contract execution when BC Smart Contract is used.

BC cannot be easily tampered with. Activities like signed by validating members, encrypted, hashed and time stamped, give rise to inherently tamper-proof characteristics. For the same reasons, any malicious activity is easily and readily detected by members. Internet connected people and helped in smooth flow of information. BC is predicted to deliver smooth flow of assets.

7.12 ADVANTAGES OF SMART CONTRACTS

- The implementation of distributed shared ledger with embedded Smart Contracts will improve cost efficiency, compliance, accountability, time savings, and enhanced privacy.
- The trusted shared distributed ledger model eliminates the centralized agency required to monitor and/or facilitate the transaction.
- Third-party authentication for transactions is eliminated.
- Transactions can be tracked without the need for centralized or third-party agency.
- Single point of failure is a concern in centralized system. This is eliminated by the distributed ledger architecture of Smart Contracts.

7.13 APPLICATIONS OF SMART CONTRACT

Smart Contracts can be considered in applications that need to fulfill regulatory compliance, product traceability, service management, fraud counterfeiting, and so on. Hence, the areas of applications could be food, energy distribution, pharmaceuticals, travel services, transportation management, aviation, financial services, asset management, B2B transfers, central clearing for derivatives, collateral and guaranteed lending, crowd funding, and so on.

7.14 INTERNET OF THINGS

Artificial Intelligence, BCT, IoT are catalysts of digital transformation. IoT is one of the disruptive technologies (Harrington, 2018) which changes the life of everyone into a smart life, whether it is people, city, buildings, transport, environment, or industry. Hence, it is also called a world-changing technology. The simplest definition of IoT is connection and communication of connected things for exchange of data (Woodfollow, 2018). The catalysts for IoT growth are communication technologies, sensors, Radio-frequency identification (RFID) devices, internet protocols (Al-Fuqaha et al., 2015), artificial intelligence, and BCT. The next phase of IoT is visualized to create a wave in intelligent decision-making by integrating diverse technologies with physical objects called intelligent devices (Al-Fuqaha et al., 2015).

7.14.1   DEFINITIONS FOR IOT

There are many definitions for IoTs. IoT is a technology to facilitate physical objects on the Internet to share information and to coordinate decisions; thereby the objects become smart objects. These smart objects make the most of the computing and communication technologies. These smart objects contribute to both the domain-specific solutions and domain-independent services (Al-Fuqaha et al., 2015).

Cisco's definition of IoT is that it is the network of physical objects connected over Internet. These physical objects are made of embedded technology which enables them to act on their internal states and also to communicate with external environment. IoT changes when, where, who, and how decisions are made in the connected world (Cisco, 2014).

While laying groundwork for RFID, Kevin Jashton foresaw that computers are to be empowered to collect data and act on the data autonomously without human intervention. The RFID and sensor devices interconnected on the Internet will play an important role in this transformation (Kevinjashton, 2009).

7.14.2   IOT DEVICES

By and large, IoT devices are miniaturized microcontroller kind of systems with default features necessary for communication over network. However, little more sophisticated IoT devices have optional features for sensing, activation, data acquisition, and minimalistic storage and computation (Zennaro, 2017).

Thus, RFID devices are the first devices that came with some memory and capabilities which helped pave way for improving supply chain, objects tracking, and so on. Later, wireless sensor networks changed the life in natural calamity recovery, industrial safety and security, national safety patrolling, agriculture, process control systems management, and so on. These devices and sensors are still part of the IoT devices. But the way, they are getting utilized has changed. Any device including our mobile handsets, wearable gadgets, which can get connected on Internet with identity and communicate and act on information, is an IoT device. Thus, IoT devices in the world are expected to cross 20 billion by 2020.

The perception on IoT devices is to exchange data on what it is meant for. Thus, these devices have some amount of intelligence and processing capabilities. Today's wearable devices are a kind of IoT devices that personalize health monitoring. IoT cameras help monitor security at the places it is installed, send the pictures over Internet. Automobile industry not only uses lots of IoT devices to increase driver and passenger experience but also monitors the vehicle health over Internet. Household appliances are becoming smarter and homes are becoming smart homes with IoT-enabled devices.

7.14.3   IOT ECOSYSTEM

Essentially an IoT ecosystem requires four components at the minimum, that is, *things* (the IoT devices identifiable with IP address and communicate without human intervention on the connected network), *network Infrastructure*, *Gateway* (Intermediary between IoT device and cloud controlling the communication aspects), and *CloudSetup* (where the IoT device data are stored and also act as computing and service center) at the minimum (Banafa, 2017). Simplest functions of IoT ecosystems are identification, sensing, communication, computation, and services (Al-Fuqaha et al., 2015). A typical IoT system is depicted in Figure 7.4.

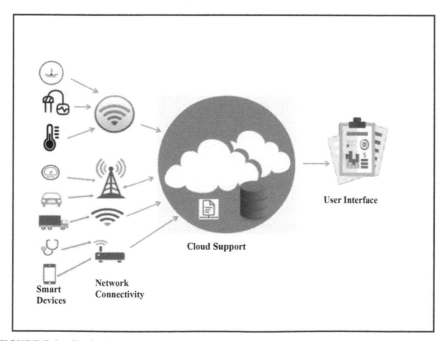

FIGURE 7.4   Typical IoT ecosystem.

The characteristics of an IoT ecosystem, as stated by ITU are inter-connectivity, heterogeneity of the devices, dynamic state changes of the devices, and interconnection of a huge number of devices (Zennaro, 2017).

7.14.4 GROWTH AND ROADMAP OF IOT

According to Forrestor Survey, industrial adoption of IoT stands at 45% and expected to reach 67% adoption in another 12 months, that is, by mid-2019 (Louis Columbus, 2018). Forbes News project that the spending on IoT in various sectors will increase two to fivefold from 2015 to 2020. The broad business segments utilizing on IoT are manufacturing, transportation and logistics, utilities, B2C, healthcare, process industry, retail, energy and natural resources, government, insurance, and so on (Louis Columbus, 2018).

With surveillance devices, vertical market applications, such as health-care, transport, food safety, document management, and so on, could benefit with cost reduction in their business. Then came the devices with ability to locate geolocation and receive signals, enabling ubiquitous computing. Now, with the miniaturization, power efficient devices and wide spectrum availability, ability to monitor distant objects, the physical world could be controlled via webcreating smart life. With advent of AI, software agents and advanced sensor fusion will take the world to next generation of smarter life. We have started hearing of vehicles working as reporter, recruiter, and so on, is an example of IoT driven by AI.

7.14.5 IOT APPLICATIONS

IoT will set a new arena of services that will not only improve the quality of life of consumers but also the productivity of enterprises (GSMA, 2014), which in turn improves the world's economy. AI–IoT and IoT–BCT solutions enhance customer experience to the extent that quality of life cannot be reverted and hence these technologies are called disrup-tive technologies. GSMA presents a viewpoint as in Figure 7.5 mapping various applications of IoT and also maps those applications in terms of their requirement of connectivity bandwidth and geomobility of the device (GSMA, 2014).

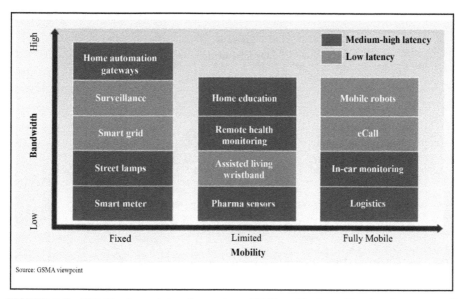

FIGURE 7.5 GSMA viewpoint and mapping of IoT applications. Reprinted courtesy of GMSA, 2014.
Source: https://www.gsma.com/iot/wp-content/uploads/2014/08/cl_iot_wp_07_14.pdf page 12

7.14.6 MANUFACTURING INDUSTRY

IoT is also considered to be the catalyst of the *Fourth Industrial Revolution* (Industry 4.0) and has triggered technological changes in various industries. Thus, IIoT provides the potential to reduce manufacturing cost, increase manufacturing efficiency and product quality. IIoT is expected to facilitate new business models with application integration.

Use IoT in manufacturing yields optimization in production, maintenance, inventory management, and energy management. Automation in production is efficient with IoT systems. Further, people are well connected to the processes with right information; hence, data acquired by IoT devices are analyzed to improve the efficiency of the manufacturing unit. Rather, few roles get rolled out into new roles in IoT-enabled manufacturing. Overall IoT systems help achieve intelligent manufacturing. The Institute for Information Industry, Taiwan, in its study on IoT solutions and development has published its perception on IoT-enabled manufacturing as in Figure 7.6 (Feng, 2015).

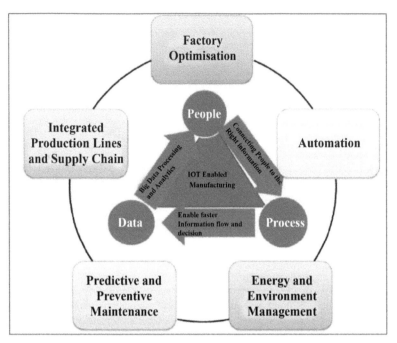

FIGURE 7.6   IoT-enabled manufacturing.
Source: Feng (2015).

7.14.7 SMART LIFE

Smart life is enabled by smart city, smart traffic, smart car, smart home, and so on. In SmartCity ecosystem, IoT-based traffic, utilities, lighting, environmental monitoring, and public safety have important contributions.

Health sector is the biggest beneficiary toward health monitoring, emergency response at home, nursing and home monitoring, and navigation system for blind people, and so on. Wearable IoT and IoT cameras enhance the quality of life and create a smart home. In smart home, many of the appliances like garage opening, coffee maker, air-conditioner, lights, and security and surveillance camera can be managed by able people efficiently even while away from home or on the way to home, as these can be connected by devices. It is all about anytime, anywhere, anything!

Smart automobiles, improve the efficiency of their vehicle and assist in maintenance of the vehicles. In a larger outlook, IoT–artificial intelligence combination help achieve traffic management, autonomous vehicles in private colonies, energy management, and smart grids.

7.14.8   OTHER APPLICATIONS

Lopez Research LLC (2013) describes the major benefits of IoT with three C's, namely, communication, control, and cost savings. Thus, IoT will help businesses by harnessing intelligence from devices to achieve the goal of improved operations and customer satisfaction. When used for personal efficiency, life becomes smarter. IoT can also change life with better and faster information in public safety, transportation, and healthcare. In hospitals, IoT devices when attached to equipment such as wheelchair, bed, trolley, nursing apparatus, and so on can help track these items and improves efficiency and utilization. Supply chain and logistics Industry is benefitted by GPS-enabled assets, such as vehicles, parcels which are useful in tracking. Control and automation industry can set baseline values and get alerts on the status of the plants, furnaces, equipment, and remotely control them too. Smart energy meters help optimize energy utilization as these can be read remotely (Lopez Research LLC, 2013). Figure 7.7 is an example of select IoT applications.

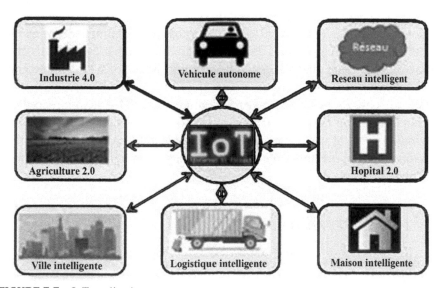

FIGURE 7.7   IoT applications.

Agriculture is another area where innovative solutions on climate control and crop health can improve the yield. Also, sensors at plants and crop fields can help the farmer monitor crops from being away.

On e-governance, social security, insurance sector, no doubt these disruptive technologies play a major role in enhancing services in terms of timely services and right services.

In education sector, fundamental services like monitoring the presence of students and faculty in the premises, not necessarily in the classroom can be managed with IoT systems. This improves the productivity of the stakeholders like students and faculty.

In democratic countries, voting systems can become efficient with smart Aadhar like card and help reduce the expenses in national elections.

Smart banking is a reality with IoT, which we practice with multiple devices, such as ATM, mobile, PoS machines, credit/debit cards, which are also the devices on the Internet.

7.14.9 ISSUES IN IOT ENVIRONMENT

7.14.9.1 DATA AND DATA HANDLING

Even if we take a specific IoT use case, each of the number of devices connected is meant to generate and act on data. These IoT devices are generally some kind of event driven for this purpose. Depending on the event schedule, each device generates data packets/transaction per event. Thus, the volume of data generated is so huge resulting in issues of data storage and data handling.

Take the case of wearable device, scheduled to send the health record of a person throughout the day and at an interval of say 10 min. If each record is saying 40 bytes of various readings of the body, there will be 144 records and 5760 bytes/day/person. This storage obviously has to be in cloud linked to the app.

7.14.9.2 SECURITY AND PRIVACY

According to a report by Online Trust Alliance for US Department of Commerce, 47% of consumers' rate security and privacy is the biggest concern to adapt IoT solutions (Kh, 2018). The security threats to computer systems are at large, with IoT devices connected to it (Suresh et al., 2014).

IoT components, namely, RFID, WSN (wireless sensor network), and Cloud, are vulnerable to attacks (Gubbi et al., 2013). Lee and Lee (2015) report that 70% of the most commonly used IoT devices in the market are vulnerable to attacks as these devices lack in cryptographic techniques, have insecure APIs, fragile software security, and inadequate authorization methods. Thus, security and privacy of the devices and users is the gravest concern in the IoT architecture. Not only those, the IoT devices are miniature in nature, have limited capability, mostly operate with limited power support, and have IP addresses. Hence these are vulnerable to be hacked, leading to doubt on reliability and dependability of the data.

The issue is very serious in nature with use of cases like smart vehicular network, industrial IoT. Any interruption or misinformation causes huge loss of unimaginable size.

7.14.9.3   NETWORK SPEED

The data from IoT devices are generally real-time. Hence, such IoT network becomes a real-time control application. However, the processing of the data and computation happens at a cloud or server. Data collection to corrective action becomes critical that the network speed is very important and critical. A scenario of driverless vehicles can be imagined for this purpose.

7.14.9.4   DDOS ATTACK

This is very common in IoT network. Lack of standards for the authorization and authentication of edge devices in the IoT network is a weak scenario. Further, Most IoT devices with simple processors and Os and sophisticated security approaches of the day are a challenge to be incorporated in the device (Banafa, 2017).

7.14.9.5   HETEROGENEITY OF DEVICES AND COMPLEXITY

Each IoT devices in a network is meant for a different purpose. These could be from different manufacturers with different operating specifications. Hence, integration is the issue. Not only this, the heterogeneity could prevail over the data packet standards and communication protocols too. Thus, the design of the network and application is a complex process. Technology changes by Moore's law time frame cause scalability and compatibility issues in maintenance of the setup.

7.14.9.6   ADDRESSABILITY OF IOT DEVICES

With billions of growths of devices, each identifiable on the global network, this was an issue with IPv4 and now resolved with IPv6.

7.14.9.7   ETHICAL AND LEGAL ISSUES

Many devices are being manufactured in different parts of the world, used globally. Newer technology, rapid and radical applications of IoT developed by enterprising techies, luring people to use, possibly compromising on the ethical and legal issues in select cases. IoT devices generally lack service agreements regarding personal level data and issues. There are limited guidelines for lifecycle maintenance and disposal of IoT devices is a concern (Banafa, 2017) and e-waste management. In use of cases, such as weather monitoring, chemical plants, the devices are planted at such remote locations that accessing them after deployment is almost impossible.

7.14.9.8   AUTHENTICITY OF IOT TRANSACTIONS

Most of the IoT devices are used in real-time scenario, which are connected over network and vulnerable for attack. Spoofing can happen. Hence, the authenticity of transaction is important. We can imagine a driverless car being hijacked by a spoofer. In some critical applications, like energy transactions, BCT could come to help verify the authenticity of transaction.

7.15   IOT SECURITY ISSUES

The physical world is becoming largely interconnected creating a smart intelligent world. But in the current state, this also endangers with security and privacy issues, as the interconnecting IoT devices are highly insecure (Woodfollow, 2018). Hackers have hacked implanted cardiac devices, driverless cars causing panic. Figure 7.8 is an example of a likely heterogeneous population of IoT devices.

FIGURE 7.8   Heterogeneous population of IoT devices.

Source: Reprinted courtesy of Coppol 2018.

IoT devices generate, process, and exchange large transactions of micro-sized data. Most of these data are safety critical from sensor networks, privacy-sensitive data from healthcare or smart home gadgets like the case may be of IoT systems integration. And hence, these IoT devices are alluring target for cyber-attacks (Sicari et al., 2015).

Many of the IoT devices, especially sensors and RFID devices, on the network are low energy and light weight, meaning their energy and processing capabilities are meant to meet only the core functions. The manageability of the resources to extend security and privacy functions is a challenge to be addressed. Further conventional methods of information and communication security are mostly centralized frameworks and not resource efficient on IoT ecosystem. IoT ecosystem is a highly distributed network environment, forcing the industry to mature on distributed security frameworks (Roman et al., 2013).

IoT security is more complicated, also because the devices use simple CPUs and operating systems with limited resource management capability to support sophisticated security approaches (Banafa, 2017; Al-Fuqaha et al., 2015). Woodfollow (2018) believes that security flaws happen due to either or all of the three factors, that is, authentication, connection, and transaction. The IoT ecosystem should address these factors.

Lack of common standard is a reason for security issues in IoT systems. Without standards, guaranteeing security and privacy in heterogeneous IoT network is difficult. Distribution of keys amongst IoT devices is still an open problem leading to security and privacy issues (Al-Fuqaha et al., 2015; Ishaq et al., 2013).

7.15.1   COMMON ATTACKS IN SMART-HOME ENVIRONMENT

In smart-home environment, *Distributed Denial of Service (DDoS) attack* and *linking attack* are most common (Dorri et al., 2017). *Denial-of-sleep attacks* drain batteries and imitate the nodes. On October 21st, 2016, a massive DDoS attack crippled most popular server and services, such as Twitter, Netflix, PayPal, and NYTimes, was found to be because of the CCTVs and smart-home devices hijacked by Mirai Malware and used against those servers (Banafa, 2017).

DDoS attack: In this case, the attacker infects a number of IoT devices in the network forcing them to communicate with a particular target node with requests that it cannot respond.

Linking attack: This is a case where the attacker manages to find the real-world identity of a user. The attacker identifies the link between transactions using the same public key. Hence, the name linking attack. This attack endangers privacy.

7.15.2   IOT SECURITY SOLUTIONS

Al-Fuqaha et al. (2015), in the survey report, bring out the following findings:

- Security to be addressed by IoT systems rather than IoT devices.
- The IoT systems may incorporate validity checks, authentication, data verification, data encryption, and so on for enhancing security.
- IoT applications to be developed with better code development standards and also these applications to be stable, resilient, and trustworthy.
- Interoperability standards for applications and network integration is essential, else, threats will be proportionately increasing with every device getting added to the network.

BCT is suggested as a solution for enhancing security, privacy in IoT by many researchers. Accessing data from IoT devices in the BCT–IoT environment requires one to pass through an additional layer of security, which is boosted by robust encryption standards. Further the distributed environment alleviates the single point failure concern of Centralized system (Gupta, 2018; Banafa, 2017; Dorri et al., 2017; Brody and Pureswaran, 2014; Zhang and Wen, 2017; Palattella et al., 2016).

Dorri et al. (2017) have suggested a BC-based smart-home framework with protection against linking attack and DDoS attack. Hierarchical defense is suggested by the author for DDoS attack. Smart-home system consists of IoT-based smart appliances and a specialized node called "miner." This miner node is always online and is a high resource device preserving a private secure BC. This device is responsible for handling all the communications amongst the smart appliances inside the home and also external to the smart-home system. This kind of BC–IoT-based smart home guarantees confidentiality, integrity, and availability and thus enhancing security. Linking attack protection can be enhanced, if each smart appliance data are handled by a unique key. Then, a ledger is created with device versus PK. This ledger is referred for each transaction by the miner (Dorri et al., 2017).

7.15.3   SYNERGY OF IOT AND BCT

Ubiquitous connectivity, IP-based networking, smart devices, technological surge in data analytics, and the convenience of Cloud Computing have brought IoT closer to the world and the world has become a smarter world. IoT ecosystems face many challenges as the devices connected and their potential is exponentially growing.

One important aspect of business is trust. BC takes away the trust from centralized intermediaries and thrusts upon technology. In doing so, enhances the security of transactions. The Smart Contracts of BC technology which are small embedded programs, seamlessly manage verification, identification, and validation, thus providing secured IoT transactions. The Smart Contracts regulate all the activity between consumer–device and device–device communications that are required for the transaction to be untamperable. Thus, BCT not only can unlock the potential new use cases but also significant return on investment from these new business use cases (VB Staff, 2018).

Although IoT and BCT are yet to mature, the exciting vision of future possibilities that the world would see in the way business is done and new business is created is not only disruptive but also delivers synergy in smart life and business (Woodfollow, 2018). Merging of IoT–BCT technologies can ensure provable, secure and immutable way of logging the data processing carried out by "smart" machines/devices. Further, the secured interaction between the connected devices enables autonomous decision making (Woodfollow, 2018). BC may still be considered as regulatory and economic risk, as in the case of cryptocurrencies, it is a revolutionizing transaction processing business.

7.15.4   IBM WAY OF LOOKING AT IOT DEVICE INDEPENDENCE (BRODY AND PURESWARAN, 2014)

IBM foresees that in future, solutions with features like drastically low cost, privacy best or autonomous solutions are going to be favorites, and these solutions are feasible only with IoT as driver. The business models for the future must adapt a highly efficient digital economy and deliver collaborative value with new disruptive products. The pyramid of digital success as suggested by IBM is shown in Figure 7.9a. Sooner, these design principles will transform the connectivity of devices as in Figure 7.9b. Since 2005, the design of IoT networks was ruled by closed-centralized model. Now the shift is toward open-centralized cloud model. By 2025, the IoT network design will only be open-access and distributed cloud model.

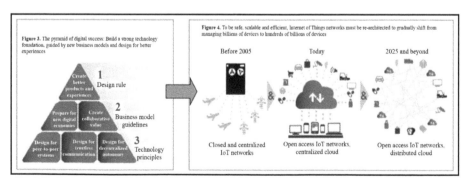

FIGURE 7.9   (a) Pyramid of digital success and (b) architectural design transition of IoT systems.
Source: Brody and Pureswaran (2014). Reprinted with permission from IBM Institute for Business Value.

In the case of centralized system, the server plays a pivotal role in functions like brokering messages to the members, arbitration, authorization, file management, and file transfer. The future open-access decentralized network will have the following three characteristics to be satisfied (Brody and Pureswaran, 2014):

Trustless P2P messaging in BCT-IoT environment

Secure data sharing in distributed environment

Scalability of devices and device coordination

Afterall, IoT devices are in billions, that is, huge number of devices possibly in every ecosystem. Hence, some form of validation and consensus will eliminate the few malicious devices, if any. Thus IoT–BCT is the solution, with the above features, providing synergy for the solutions (Brody and Pureswaran, 2014).

The future is about decentralized framework for transaction processing eliminating central agency. So is the case with IoT devices in transaction world. Transaction world sooner will be completely ruled by IoT. Decentralized IoT–BCT architecture combined with the power of AI is anticipated to conclude with "Internet of Decentralized, Autonomous Things." This applied to transaction world will be the democratized digital world. The democratized digital world will change the characteristics of the physical assets to be liquid, personalized, and digitally efficient. The assets will be digitalized for transaction. Further, IoT will be a compelling business strategy than a technical curiosity. IoT–BCT solutions can bring in liquidity to the industries that use them, with greater productivity and profitability of unimagined level, similar to that of financial markets.

7.16   BCT STRENGTHS FOR IOT

Decentralization, fault tolerance, attack resistance, collusion resistance, authentication, trust, immutable, traceability, consensus, Smart Contract, privacy, reliability are the BCT strengths for IoT.

7.17   FOG COMPUTING, BIG DATA, AND CLOUD

The billions of devices are useful only if the data from these are processed. These heterogenous devices generate data in variety, velocity, and veracity, mandating the processing be better with Big Data approaches. Of course, these tiny devices need a location where data captured are stored and processed and commands are returned. Cloud is the default model in centralized network. Of course, these days, distributed IoT network is being discussed and hence the cloud probably gets replaced with edge computing or fog computing (Al-Fuqaha et al., 2015). For fog computing, being at proximity to the devices, the services are expected to be faster and may also address the security and privacy issues locally. Fog architecture for IoT will be a scalable model with improved performance.

7.18   E-BUSINESS ARCHITECTURE

Zhang and Wen (2017) proposed BC to enhance IoT Security. Zhang believes that traditional e-business models do not fit well with e-business on IoT platform. IoT platform for business is imminent because IoT can make the environment cognitive by enabling the status of the goods to be in its product cycle, say manufacturing, transportation, consumption, and so on. Zhang further proposes a Smart Contract based e-business architecture to trade any smart property which is digitalized and traded. The BC Smart Contract along with cryptocurrency or IoT coin addresses security, transparency, traceability, and so on issues of business.

7.18.1   5G TECHNOLOGY

Consumer IoT (CIoT) which gives smartlife to people and IIoT which focuses on operational technology and information technology are being differentiated with the underlying technologies and business model to give better performance in their field of application. Although CIoT and IIoT have got popularized, next are the autonomous systems which are most demanding in terms of the connectivity, computing capability, integration, and reduced latency. 5G technology is expected to address these demands, with its unified interconnection framework and provide a seamless connection of "things" on the Internet. The implications of 5G on business models and technology will have greater implications by 2025 (Palattella et al., 2016).

7.18.2   IOTA

IOTA is also a distributed ledger like BC but uses different technology called "Tangle" based on directed acyclic graphs for validation of transactions in the chain. Consensus is inbuilt and intrinsic as part of IOTA, that every member can participate in consensus. In addition, every transaction will have link to the previous two blocks. IOTA is said to be lightweight that BCT and uses quantum cryptography. IOTA is highly scalable and resource requirements are low. IOTA is meant for tiny devices to participate in transactions, making it suitable for IoT. Possibly over the days to come, IOTA could give a fillip to IoT much easily than BCT (IOTA, 2018).

KEYWORDS

- **internet of things**
- **blockchain technology**
- **industrial IoT**
- **blockchain**
- **Industry 4.0**

REFERENCES

Al-Fuqaha, A.; et al. Internet of Things: A Survey on Enabling Technologies, Protocols, and Applications. *IEEE Commun. Surv. Tutor.* **2015,** *17* (4), 2347–2376 (accessed on Feb. 19, 2019).

Back, A.; Corallo, M.; Dashjr, L.; Friedenback, M.; Maxwell, G.; Miller, A.; Poelstra, A.; Timon, J.; Wuille, P. *Enabling Blockchain Innovations with Pegged Sidechains*, 2014. Retrieved from http://www.blockstream.com/sidechains.pdf.

Banafa, A. IOT and Blockchain Convergenge Benefits and Challenges. *IEEE Newslett.* **2017.** https://iot.ieee.org/newsletter/january-2017/iot-and-blockchain-convergence-benefits-and-challenges.html (accessed on Feb. 19, 2019).

Brody, P; Pureswaran, V. *Device Democracy: Saving the Future of the Internet of Things.* IBM Institute for Business Value. https://www.ibm.com/downloads/cas/Y5ONA8EV (accessed on March 4, 2019).

Buterin, V. *On Public and Private Blockchains*. Ethereum Blog, 2015. Retrieved from https://blog.ethereum.org/2015/08/07/on-public-and-private-blockchains/.

Dorri, A.; Kanhere, S. S.; Jurdak, R.; Gauravaram, P. Blockchain for IoT Security and Privacy: The Case Study of a Smart Home. In *2017 IEEE International Conference on Pervasive Computing and Communications Workshops (PerCom Workshops)*, IEEE, March 2017; pp. 618–623.

Feng, M-W. Internet of things (IOT) Trend and Solution Development in Taiwan. 2015. https://www.slideshare.net/agencedunumerique/iot-trend-and-solution-development-in-taiwan

Ford, B. *Blockchain for Beginners*. 2016. http://bford.info/log/2016/1102-cybsec-blockchain.pdf (accessed on Feb. 27, 2019).

Gatteschi, V.; Lamberti, F.; Chiara Pranteda, D.; Santamaría V. Theme Article: Financial Technologies and Applications—To Blockchain or Not to Blockchain: That Is the Question. *IT Professional*; IEEE Computer Society, 1520-9202/18/March–April 2018.

Gubbi, J.; Rajkumar, B.; Slaven, M.; Marimuthu, P. Internet of Things (IoT): A Vision, Architectural Elements, and Future Directions. *Fut. Gener. Comput. Syst.* **2013**, *29*, 1645–1660.

Gupta, M. *Blockchain for Dummies*, 2nd IBM Limited Edition; John Wiley & Sons: Hoboken, NJ, 2018.

Harrington, L. 5—*Disruptive Technologies Shaping Our Future*, Nov. 20, 2018. https://www. iotforall.com/5-disruptive-technologies-shaping-our-future/ (accessed on Feb. 20, 2019).

Ishaq, I.; Carels, D.; Teklemariam, G. K.; Hoebeke, J.; Abeele, F. V. D.; Poorter, E. D.; Moerman, I.; Demeester, P. IETF Standardization in the Field of the Internet of Things (IoT): A Survey. *J. Sensor Actuat. Netw.* **2013**, *2*, 235–287.

Johan, Ä.; Jansson, M. *Investigation of Blockchain Applicability to Internet of Things within Supply Chains*. Master's Thesis, Uppsala Universitet, Jun. 2018.

Kh, R. *Patch Management Is the Catalyst for Growth in the IoT Industry*, Jun. 13, 2018. https:// datafloq.com/read/patch-management-catalyst-growth-internet-things/5106 (accessed on Feb. 17, 2019).

Lee, I.; Lee, K. The Internet of Things (IoT): Applications, Investments, and Challenges for Enterprises. *Bus. Horizons* **2015**, *58* (4), 431–440.

Lopez Research LLC. *An Introduction to the Internet of Things (IoT)*; Lopez Research LLC, 2013 (accessed on Feb. 17, 2019).

Louis Columbus. *10 Charts That Will Challenge Your Perspective of IoT's Growth*; Forbes, June 2018. https://www.forbes.com/sites/louiscolumbus/2018/06/06/10-charts-that-will-challenge-your-perspective-of-iots-growth/#41c3f2113ecc (accessed on Feb. 20, 2019).

Miraz, M. H.; Ali, M. Applications of Blockchain Technology beyond Cryptocurrency. *arXiv* **2018**, arXiv:1801.03528.

Palattella, M. R.; Dohler, M.; Grieco, A.; Rizzo, G.; Torsner, J.; Engel, T.; Ladid, L. Internet of Things in the 5G Era: Enablers, Architecture, and Business Models. *IEEE J. Select. Areas Commun.* **2016**, *34* (3), 510–527.

Pilkington, M. 11 Blockchain Technology: Principles and Applications. *Res. Handb. Dig. Transform.* **2016**, 225.

Raja. *Blockchain Infographic: Growth, Use Cases & Facts by Raja*, May 14, 2018. https:// www.dotcominfoway.com/blog/growth-and-facts-of-blockchain-technology.

Rennock, M. J. W.; Cohn, A.; Butcher, J. R. Blockchain Technology and Regulatory Investigations. *J. Litigat.* **2018**.

Roman, R.; Zhou, J.; Lopez, J. On the Features and Challenges of Security and Privacy in Distributed Internet of Things. *Comput. Netw.* **2013**, *57* (10), 2266–2279.

Sicari, S.; Rizzardi, A.; Grieco, L. A.; Coen-Porisini, A. Security, Privacy and Trust in Internet of Things: The Road Ahead. *Comput. Netw.* **2015**, *76*, 146–164.

Sultan, K.; Ruhi, U.; Lakhani, R. Conceptualizing Blockchains: Characteristics & Applications. *arXiv* **2018**, preprint arXiv:1806.03693.

Suresh, P.; Daniel, J. V.; Parthasarathy, V.; Aswathy, R. H. *A State-of-the-Art Review on the Internet of Things (IoT) History, Technology and Fields of Deployment*; IEEE, 2014.

Swanson, T. *Great Chain of Numbers: A Guide to Smart Contracts, Smart Property and Trustless Asset Management*, 2014, CC4.0; p 11.

Swanson, T. *Consensus-as-a-Service: A Brief Report on the Emergence of Permissioned, Distributed Ledger Systems*. Working Paper. April 6, 2015. Retrieved from http://www. ofnumbers.com/wp-content/uploads/2015/04/Permissioned-distributed-ledgers.pdf.

Wang, H.; Zheng, Z.; Xie, S.; Dai, H. N.; Chen, X. Blockchain Challenges and Opportunities: A Survey. *Int. J. Web Grid Serv.* **2018,** *14*, 352. 10.1504/IJWGS.2018.10016848.

Woodfollow, J. *Blockchain of Things—Cool Things Happen When IoT & Distributed Ledger Tech Collide*, Apr. 2018. https://medium.com/trivial-co/blockchain-of-things-cool-things-happen-when-iot-distributed-ledger-tech-collide-3784dc62cc7b (accessed on Dec. 19, 2019).

Wright, A.; De Filippi, P. *Decentralized Blockchain Technology and the Rise of Lex Cryptographia*; Mar. 10, 2015. Retrieved from http://ssrn.com/abstract=2580664.

Yli-Huumo, J.; et al. Where Is Current Research on Blockchain Technology? A Systematic Review. *PLoS One* **2016,** *11* (10), e0163477.

Zennaro, M. *Introduction to IOT*, 2017. https://www.itu.int/en/ITU-D/Regional-Presence/AsiaPacific/SiteAssets/Pages/Events/2017/Nov_IOT/NBTC%E2%80%93ITU-IoT/Session%201%20IntroIoTMZ-new%20template.pdf (International Telecommunication Union), Telecommunications/ICT4D Lab, The Abdus Salam International Centre for Theoretical Physics: Trieste, Italy (accessed on Feb. 20, 2019).

Zhang, Y.; Wen, J. The IoT Electric Business Model: Using Blockchain Technology for the Internet of Things. *Peer-to-Peer Netw. Appl.* **2017,** *10* (4), 983–994.

Zheng, Z.; et al. An Overview of Blockchain Technology: Architecture, Consensus, and Future Trends. In *Big Data (Big Data Congress), 2017 IEEE International Congress on.* IEEE, 2017.

WEB REFERENCES

Cisco. http://www.cisco.com/web/solutions/trends/iot/overview.html, 2014 (accessed Jan. 21, 2014).

Kevinjashton. http://kevinjashton.com/2009/06/22/the-internet-of-things/, 2009.

GSMA. 2014. www.gsma.com/connectedliving.

IOTA. 2018. https://www.iota.org/get-started/what-is-iota (accessed on Feb. 20, 2019).

https://gendal.me/2015/02/10/a-simple-model-for-smart-contracts/ (accessed on Feb. 28, 2019).

https://medium.com/@neocapita/breaking-down-blockchain-6301081cb525 (accessed on Feb. 25, 2019).

https://venturebeat.com/2018/03/22/blockchain-iot-and-the-future-of-seamless-transactions-vb-live/ (accessed on Feb. 17, 2019).

https://web.iii.org.tw.

https://www.mckinsey.com/business-functions/digital-mckinsey/our-insights/blockchain-beyond-the-hype-what-is-the-strategic-business-value.

STATISTA. https://blogs-images.forbes.com/louiscolumbus/files/2018/06/Spending-on-IoT-by-vertical.jpg (accessed on Feb. 17, 2019).

CHAPTER 8

Blockchain Internet of Things: Security and Privacy

KAVITA ARORA

Manav Rachna International Institute of Research and Studies, Faridabad 121001, Haryana, India

*Corresponding author. E-mail: kavita.fca@mriu.edu.in

ABSTRACT

In today's world, the technologies like Blockchain and Internet of Things (IoT) are considered to be the inevitable world-changing technologies. In today's age where the revolution of IoT is laying foundations for a world to bring it to possibility for interconnection and automation of routine tasks that further requires certain enhanced features like security, privacy of data, easy deployment, robustness, and so forth, can possibly be achieved by a technology called Blockchain, which is born with a cryptocurrency popularly known as Bitcoin. So, if we merge them together, it will give rise to a network of interconnected devices with capability of interaction of smart devices and decision-making even without human intervention. Keeping this in view, without exaggerating, Blockchain technology might prove to be like a silver bullet to the IoT world. If we think of any failure in IoT world, it comes with a threat exposure to the safety of the whole system. Here comes the role of Blockchain, which proves out to be a boon for managing the data across all IoT devices as it has the cushion of supplementary layer of security with many vigorous encryption standards.

In this chapter, we will discuss about the issues such as security and privacy related to IoT, which on the one hand, is confronting rampant incubation in the fields of research and industry, and on the other hand still has to endure with finding the solution of the issues mentioned earlier. Herein the Blockchain approach that underpins the cryptocurrency can further be used to dispense security and privacy to IoT.

8.1   INTRODUCTION TO BLOCKCHAIN TECHNOLOGY

The concept of blockchain was first of all given by Satoshi Nakamoto in 2008. Blockchain that underpins Bitcoin is contemplated to be one of the most propitious technologies of this era. The technology named as block-chain is neither an organization nor an application, instead it is an absolutely novice method for the documentation of the data on the Internet. Even though blockchain is in its juvenile stage but the pervasive development is just not very far away.

A blockchain works on the concept of storing the data digitally. The data transpires in the form of blocks and the blocks are chained together to form the data ineradicable. It means that as long as the blocks are chained together, the stored data cannot be altered at any stage. This data is all the time acces-sible to the public as and when the user wants to see it and precisely in the same manner as initially stored in the blockchain (what is Blockchain, 2018).

In other words, it is imperatively a decentralized administered account book (a database or record book) that records and transfers data in a way that is clear, shielded, auditable, and resistant to breakdown. It permits the storage of every negotiation into ineradicable form and each transaction is dispensed throughout the network among the participating nodes.

In blockchain, the term Bitcoin is assigned to electronic cash with the help of which the transactions are done in the decentralized environment from peer-to-peer using public-key cryptography.

Two main components of a blockchain:

i)   Transaction: These are the set of activities taking place by the partici-pants in the system.
ii)   Blocks: Blocks are the main building blocks as they are responsible for recording the transactions and that too in a proper sequential order and in an untampered condition. The time slot is also marked as to keep a record when did the transaction take place.

8.2   BASICS OF BLOCKCHAIN

The technology has been named as blockchain and it beholds this name on the basis of its working and the way the accumulated information is stored into blocks and links together all the blocks of the alike statistics in the form of a chain. It is this connectivity because of which it becomes almost impossible to make modifications in any information or to put in any new block in between any existing blocks. In case, anyone attempts to do so, all the trailing blocks would also need to be altered edited. Thus, every block in blockchain provides strength to the precedent block and the guarantee of the security to the complete blockchain as it ensures that no one would be able to fiddle with the stored information (Chris).

As the name suggests, blockchain is a chain of time-stamped blocks linked together with the help of cryptographic hashes. In this technology, blocks are the primary building blocks, which are the small sets of transactions taking place. Every newly added block keeps a track record of every preceding transaction taken place with the help of SHA-256 hash. As a result of this, a "chain" of blocks is created and so is the name.

As already discussed, whenever the data is inscribed in a block it cannot be changed without changing every emanating block that has been recently added and thus it is a great idea of saving the data and making it entrenched.

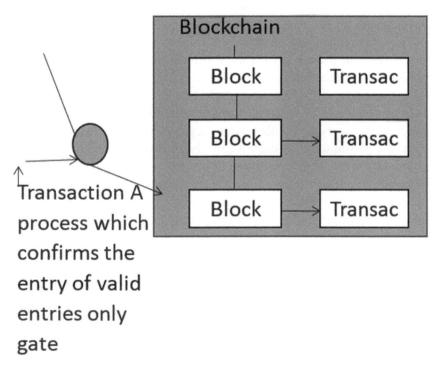

FIGURE 8.1   Structure of a blockchain.

8.3   BLOCKCHAIN BASIC FUNCTION MECHANISM

Technically, as cryptocurrency states, a block maintains the record of a single transaction taking place and a blockchain is a diary that's almost impossible to be forged. As soon as every block is concluded, it is linked to the chain and in turn creates a chain of blocks called blockchain.

If a transaction is to be added to a chain, it is to be validated by every participant node. The process of validation is done by implementing an algorithm on to the transaction. All the approved transactions are packed in a block and forwarded to every other node in the network where a new block is validated. Every consecutive block has a hash (a string of numbers and letters), which is the sole identification of the preceding block. In a blockchain, in order to tamper one block, the previous block is to be tampered and then every other block followed by, in the chain to modify it completely, it is strongly regarded as tamper resistant.

In order to understand the working of a blockchain in detail, we must learn it in the following way. Technically, to use a blockchain, first, a peer-to-peer (P2P) network is created in which all the nodes participate in creating a blockchain. Every node in the network uses two keys: a public key and a private key. In practice, the public key is used to convert the messages into cipher text, and the private key is used for the purpose of decryption of data.

Whenever a transaction is carried out, it is signed and is broadcasted to the following nodes. This process is carried out in a distinctive manner with the help of cryptography wherein it guarantees integrity of the data as the message can be decrypted with private key only and in case, if any error occurs while transmitting the data, it can't be decrypted.

In nutshell a blockchain is like a container that contains every information related to every transaction. Every transaction introduces a hash. A sequence is a set for all the transactions in the order of their occurrence. Order is very important. The hash is dependent on two components, that is, the current transaction as well as a hash of the previous transaction. Every single and minute alteration in a transaction leads to a new hash. Every node in the network validates every transaction by inspecting the hash. When the transaction gets a nod of approval by a number of nodes in the network, then only it can pop-in the block. Every block keeps a reference of the antecedent block and link together to form the blockchain.

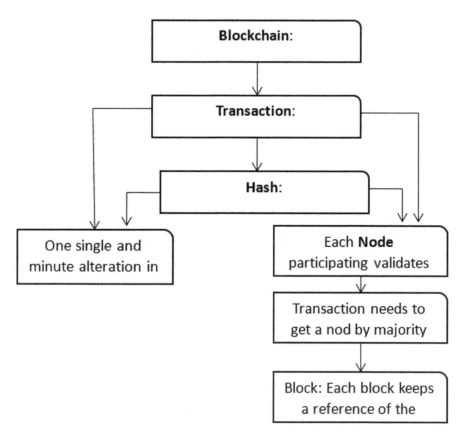

FIGURE 8.2   Working mechanism of blockchain.

A step-by-step review:

i)   A user requests for a transaction.
ii) A block representing the transaction is created.
iii) The block is broadcasted to all the nodes of the network.
iv) All the nodes validate the block and the transaction.
v)   The block is added to the chain.
vi) The transaction gets verified and executed.

8.4 KEY STRENGTHS OF BLOCKCHAIN TECHNOLOGY

In recent times, blockchain has accomplished a huge success. It has got quite distinctive characteristics like transparent P2P communication, no third-party involvement, and so on. The key strength areas of blockchain technology are:

i) **Can't be corrupted:** In blockchain, every node keeps a record of every other transaction. Whenever a new transaction is to be added, its validity is checked by every other node. Once its validity is approved by every other node, only then it is added to the ledger. This process thus promotes its transparency and hence it cannot be corrupted.

ii) **Decentralized technology:** The term decentralized means that there is no governing body or any other person looking after the framework. Rather a group of nodes are there to maintain the network to make it decentralized.

iii) **Enhanced security:** As there is no need for central authority, no one can make any changes in the network and as it also uses encryption (cryptography), it adds one more layer of security in the system.

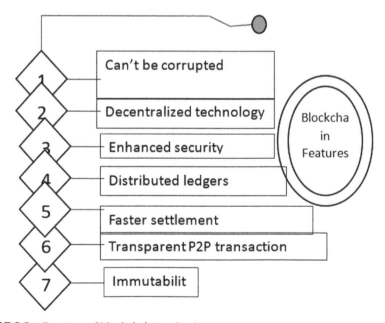

FIGURE 8.3   Features of blockchain mechanism.

iv) **Distributed ledgers:** The ledger on the network is maintained by all the other users on the system. It permits the storage of every negotiation into ineradicable records.

v) **Faster settlement:** This technology offers fast transactions in comparison to traditional banking system. This in turn helps the users to transfer the money faster with saving of time too.

vi) **Transparent P2P transaction:** P2P is one of the building blocks for the blockchain technology. A blockchain protocol operates on top of the Internet, on a P2P network of computers that all run the protocol and hold an identical copy of the ledger of transactions, enabling P2P value transactions without a middleman though machine consensus.

vii) **Immutability:** The transaction data is stored in the blockchain permanently and then can't be altered. As the system runs on algorithms, there is no chance for anyone to make modifications in the records.

Some additional advantages of using blockchain technology are:

i) It is strenuously fiddle free (tamper-proof).

ii) It is highly extensible as there is no single point of failure and attacks through distributed network nodes.

iii) It proves to be an entrenched system of records for all the stakeholders.

iv) The cost involved is less as it diminishes the involvement of manual activities taking place at administrative ends.

v) There are less chances of fraud as it shares the entries through a common immutable ledger across the network.

vi) Blockchain being decentralized in nature can survive any malicious attack and so it is less likely to break down.

8.5 APPLICATION AREAS OF BLOCKCHAIN TECHNOLOGY

i) Real estate industry

ii) Banking

iii) Healthcare

iv) Legal

v) Retail

vi) Insurance

The abovementioned discussion proves that blockchain technology isn't just a hype that would be forgotten in a few days, rather with all its advantages and applications, it is going to stay a longer as it has created another level of its impact on the web.

8.6   INTERNET OF THINGS

Internet of Things (IoT) is a computing notion that lays down the idea of connecting the physical objects or digital or mechanical objects together. For IoT to work the sensors collect the information and the Internet delivers it to the world. It is extending at a very high speed and as per the reports, the ratio of these connected devices will reach in billions by 2020–2021. Gartner, Inc. forecasted that by 2020, 20.8 billion IoT devices will be connected to the Internet (How to Secure the Internet of Things (IoT) with Blockchain, 2018).

The concept of IoT was coined by a member of the radio frequency identification (RFID) development community in 1999. IoT has been defined as, "The Internet of Things (IoT) is a system of interrelated computing devices, mechanical and digital machines, objects, animals or people that are provided with unique identifiers and the ability to transfer data over a network without requiring human-to-human or human-to-computer interaction."

It is basically a pervasive internet network of a variety of objects known as "things" and users. This technology permits the "things" and the people to couple together and to correspond and hence converting the world of physical objects into a big information system. The prototype of IoT strive for a world in which a number of routine objects would be connected and be able to communicate with the environment for the purpose of collection of information and automate the tasks.

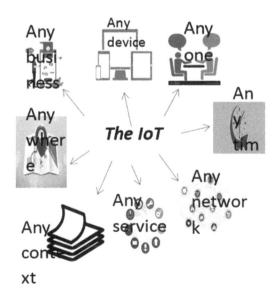

FIGURE 8.4   Application areas of IoT.

Things that are under the scope of IoT are actually a combination of hardware, software, data and services, and the "things" refer to a wide variety of devices such as electronic instruments, alarm clock, home automation systems, and so forth. In IoT, the objects that are being connected to the internet can be divided into below-mentioned categories:

i) **Things that collect information and then send it:** This task is accomplished by the sensors. As there are different kinds of sensors like temperature sensors, motion sensors, moisture sensors, air quality sensors, light sensors, they along with a connection, collect the information from the environment automatically and provide us a chance to take more intelligent decisions.

ii) **Things that receive information and then act on it:** The modern-day machines are capable of getting information and then implementing or executing just like a printer receives a command to print a document and then prints it, and our cars receive a signal from the car keys and they execute accordingly as either the doors open or locked.

iii) **Things that do both:** Such things show the actual potential of IoT wherein the things are capable of doing multiple jobs simultaneously such as collecting and sending information as well as receiving information and then to act on it.

IoT has covered a journey of various structures and architectures from being connected with a mainframe to a few objects to the present stage of connectivity of objects with cloud to the future age of interconnectivity of all the objects. Figures 8.6a, 8.6b and 8.6c illustrate the past, present, and future structures of architectures of IoT.

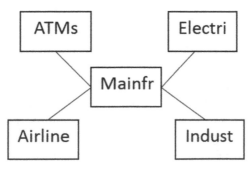

FIGURE 8.5A   Past architecture of IoT.

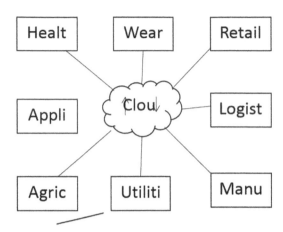

FIGURE 8.5B   Present architecture of IoT.

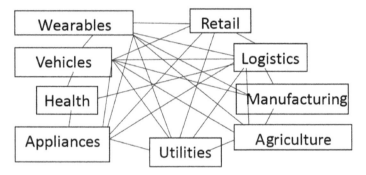

FIGURE 8.5C   Future architecture of IoT.

8.7   BLOCKCHAIN AND IOT

In today's time, IoT is almost everywhere. Within a short span of time, it has outgrown in the terms of popularity and capabilities in order to become the horizon less technology. It is said so as this technique has outstandingly amplified its capacities and coupled a lot of devices and networks in homes, workplaces, and transportation systems.

IoT is growing with another technology that has already started a serious run is blockchain. This technique, which is now almost a decade-old, is all set to dramatically reorganize the business models with the encrypted and distributed ledger, which has been designed to create tamper-proof and real-time records. Blockchain technology, which is powered by cryptocurrencies like Bitcoin, being decentralized in nature can prevent the jeopardized systems from propelling any fallacious information and scrambling the network environment.

If these two technologies are merged together, the combination would prove itself as an empirical, reliable, and persistent way for the recording of data to be processed by "smart" machines. Such a kind of network helps the machines to communicate with the environment with no or less need of any human intervention.

In the present era, if there is a failure in an IoT, it exposes multiple devices to the threat of safety and security. The origination of such security flaws generally ponders over encompassing: authentication, connection, and transaction.

Prospective risk factors for IoT devices encompass no visibility, no identity, multiple platforms, and unknown processes to upgrade the device, no automation for breach response, and many more. In the previous times, it has been observed that blockchain can be used as a weapon in order to taut the devices and systems that are based on IoT. Blockchain might prove to be a silver bullet that is most required by the IoT industry.

So, we can say that the decentralized and autonomous prowess of blockchain claim as a classic constituent of all IoT panacea. Blockchain can prove to be a boon to fix certain issuessuch as privacy, reliability, scalability, and security in IoT. We can use blockchain for tracking the linked systems and to coordinate among the devices. Blockchain being decentralized in nature would be remove single points of failure and creating a stronger symbiosis for devices to run on. In addition to all this, the cryptographic algorithms also ensure to hold the data of consumers to be highly secure and private (Stanciu, 2017).

8.8   KEY ISSUES IN IOT

The researchers have suggested that the impact of IoT is going to be enormous in the years to come. The rise in the impact and application can be judged from different available patterns in the fields of entertainment and media, agriculture, transportation, and medical care. As the current trend depicts, this influence of IoT on human life is rising rapidly and will proceed with time.

IoT has become a hot topic in the tech world and the speed with which it has started getting into the mainstream industry, the shadow of cyber-crime is following it at a higher pace, which is exploiting IoT devices and networks. The result is that the key challenges of the technology are also fast emerging.

As for IoT to function the connected devices need to take information from the user, it means there are security risks. This results in slow IoT development and its fragility toward cyberattacks, the reason being, lack of a robust centralized security model.

One of the thrust areas to deploy IoT is its security and mentioned later are the other issues for IoT infrastructure and services:

i)   As the anticipation of devices in the infrastructure is rising at a giant rate, it becomes a tough task to identify, authenticate, and secure the devices.

ii) It is quite a difficult task to employ the centralized security model and is an expensive affair too to scale, maintain, and manage.

iii) A centralized security infrastructure may prove to be a single point of failure and an unchallenging chore for DdoS attack.

iv) There may be a great difficulty in implementing the centralized infra-structure in the industrial setup as the extremity nodes are rampant across the globe.

In an IoT network, blockchain is capable enough to keep an entrenched dossier of the history of smart devices, which in turn entitles the devices to self-govern in the absence of any centralized authority. Thus, blockchain gives an opening to a progression of IoT framework, which otherwise would have been a tough nut to be cracked in absence of a blockchain. So, in every way, IoT and blockchain make a perfect pair.

It is quite apparent from the past that when at any moment of time, two corresponding technologies integrate with one another, quite a novice and potential outputs are received. Similar is the combination of block-chain and IoT where the pronounced benefit dispensed by blockchain technology is security of data, which is the most arduous challenge, faced by IoT. There is a certain other important feature of blockchain such as immutability, decentralization, and transparent data-sharing that are missing in IoT, and this is the reason as to why the Internet of Technology Community is interested in integration of the abovementioned features for data security.

Keeping the earlier discussion in mind, we can say that the blockchain technology seems to be a viable solution for all of the issues faced in the path of IoT. In the combination of IoT and blockchain, as the two go hand in hand, blockchain ensures providing a verifiable and secure method for storing the data related to the devices and the processes associated with IoT. In turn, the smart IoT devices can use this data stored on blockchain efficiently for accomplishment of various tasks.

8.9   SECURITY OF IOT WITH BLOCKCHAIN

Even though IoT is like the talk of the town nowadays still its security is known as "a doomsday scenario waiting to unfold." There are problems attached with it like exorbitant deployment failure rates and cyber threats. The security and privacy of IoT is a great challenge because of the massive scale and distributed nature of IoT networks. This situation turns out as if IoT is hard to be established because establishing benevolent IoT systems that can run in a secured, efficient, and independent manner have witnessed to be astonishingly arduous.

IoT majorly includes the devices that foster action and trade-off great extent of security and safety-critical data as well as privacy-sensitive information and thus become prone to different types of targets of various cyberattacks. In order to prevent itself from the security issues, IoT requires an extendible and dispensed security and privacy shield. The blockchain technology that foster Bitcoin, the first cryptocurrency system has got the prospects to take over the aforementioned challenges as it has distributed, secure, and private nature.

The ray of hope comes from blockchain technology as it shows promise to ease out that thrust area. As the blockchain has been discussed earlier, it throws light on a certain area regarding finding the solution for the security of IoT. One of the most important reasons of using blockchain technology is its usage as a distributed recording system. It permits to write the immutable records in a secured manner with the help of cryptography.

Thus, with blockchain every device in the network will be backed by a strong cryptography, which in turn confirms secure communication with rest of the devices and also provides privacy and security as it is the thrust area in IoT.

Blockchain +
IoT = Secure

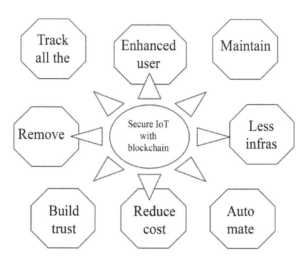

FIGURE 8.6   Features of secure IoT with blockchain.

Some major benefits of using blockchain in securing IoT are:

i)   **Track all the connected devices in the network**: IoT needs a robust centralized security model to grow in acceptance and use—something blockchain technology and the cryptographic processes driving it can offer. More and more offline objects are being converted into online assets with IoT, thereby increasing the risk of cyber threats. Businesses, therefore, need to plan strategically for the protection of not just the smartphone and desktop computer, but almost everything connected to the internet. The objects in IoT ecosystem can act analogous to the nodes in blockchain technology, thus making connected home, connected cars, and other connected devices more reliable and secure with blockchain-based IoT.

ii) **User interconnectivity is enhanced/authenticate the users and devices**: As blockchain dispense a transparent mechanism, it is helpful to the users to bring out the best utilization of the interconnectivity provided by IoT to the networks of smart devices. It also enables the users of IoT with a better connectivity with each other in many useful means and ways and discuss their issues and queries more openly to gain different insights.

iii) **Maintain data securely**: As the amount of data of data is enhancing, privacy of this data is gaining more emphasis as powerful IoT companies are tracking almost everything about the users. Lack of proper encryption and data leakage not only put individuals at risk, but also makes industry wide business models vulnerable. The combination of blockchain and IoT, on the one hand, ensures the tamper-proof, immutable, evincible, and transparent data and on the other hand the data is encrypted in a correct manner in order to safeguard the privacy of the user.

iv) **Remove single point of failure:** Blockchain can remove the single point of decision-making as it is the reason behind the failure and this is done by enabling device network to protect themselves in other ways, such as allowing devices to form group consensus about what is normal within a given network, and to quarantine any nodes that behave unusually.

v) **Build trust between IoT processes**: The trust in IoT data can be formed by enabling the five digital security primitives: availability, auditability, accountability, integrity, and confidentiality. In blockchain mechanism, the data gets stored automatically in multiple locations and is always accessible to users. For the purpose of auditability and accountability, we require a private, permission-based blockchain, which is accessible to all users who are authorized to access the network and since complete data stored on the blockchain is signed, each device is accountable for its actions.

vi) **Reduce cost by eliminating intermediaries**: Using blockchain, the scope for human intervention becomes less as the services and transactions made by IoT devices are automated.

vii) **Automated transactions**: Using blockchain, the services and transactions made by IoT devices can be automated without human interference. For instance, an autonomous car would be able to pay for a refill, parking fees or mechanic fees by enabling smart contracts that follow the "IFTTT (IF-this-THEN-that)" command.

viii) **Less infrastructure and operational costs**: Using blockchain, the infrastructure for projects can be outsourced and the cloud side of

IoT can be handled by the network. With the help of this, there is no or less requirement of any middleman and thus many costs like the legal and contractual cost, along with the maintenance, update, and supervision costs are cut down. This gives rise to abundant time and money on the side of companies to converge more on the hardware, application, and innovation side of things.

8.10 CONCLUSION

The IoT has the capacity to be a transformative force, positively impacting the lives of millions worldwide, says Bingmei Wu, Deputy Secretary-General of the China Communications Standards Association. An IoT environment is exposed to numerous issues enumerating confidentiality, privacy, and data integrity. That is why, the people working in this sector have decided to merge "security by design" technology within an environment so that IoT can triumph over the limitations. Blockchain, being one such technology, grants authenticity, non-repudiation, and integrity by default, and utilizing smart contracts, manages authorization and automation of transactions as well.

In the days to come, in place of conventional and centralized mechanism, the decentralized mechanism of blockchain will take place in order to remove the hindrances such as tampering of data, single point of failure, and interference of any third party. As in blockchain technology, the complete data across the globe is shared on the network, this will help the IoT to achieve the future structure of secured mesh networks, wherein the IoT devices will be interconnected in a secured manner and the network will be ascendable in order to support billions of devices without the need of any additional devices.

Thus, the merger of blockchain technology and IoT can prove to be a boon and a game-changing technology for the globe. This will help numerous organizations to take extra advantage of the additional features of blockchain like dependability and self-regulation. In the days to come, the world is going to experience how blockchain solutions will emerge as a ruler technology in the commercial IoT world and futuristic companies will carry forward the process of developing and implementing it themselves.

KEYWORDS

- **blockchain**
- **IoT**
- **cryptocurrency**
- **Bitcoin**
- **network**

REFERENCES

Akhtar, F.; Rehmani, M. H.; Reisslein, M. White Space: Definitional Perspectives and Their Role in Exploiting Spectrum Opportunities, *Telecommun. Policy* **2016,** *40* (4), 319–331. http://dx.doi.org/10.1016/j.telpol.2016.01.003.

Alaba, F. A.; Othman, M.; Hashem, I. A. T.; Alotaibi, F. Internet of Things Security: A Survey. *J. Netw. Comput. Appl.* **2017,** *88* (Suppl. C), 10–28. http://dx.doi.org/ 10.1016/j.jnca.2017.04.002.

Atlam, H. F.; Alenezi, A.; Alassafi, M. O.; Wills, G. B. Blockchain with Internet of Things: Benefits, Challenges, and Future Directions. *Int. J. Intell. Syst. Appl.* **2018,** *41* (10), 40–48.

Banafa, A. *A Secure Model of IoT for Blockchain. https://www.bbvaopenmind.com/en/ technology/digital-world/a-secure-model-of-iot-with-blockchain/*

Blockgeeks, *What is Blockchain Technology? A Step-by-Step Guide for Beginners.*

Chris, A. *Problems with the Internet of Things You Need to Know.*

Computer Weekly.com. *How Blockchain can Secure the IoT.*

Datafloq. *How to Secure the Internet of Things (IoT) with Blockchain.* DevOps, Using Blockchain to Advance IoT.

Eastwood, G. *4 Critical Security Challenges Facing IoT.*

Fernandez-Caramesi, T.; Fragalamas, P. *Blockchain for the Internet of Things.*

i-scoop. *Blockchain and the Internet of Things: The IoT Blockchain Opportunity and Challenge.*

Khudnev, E. Blockchain: Foundation Technology to Change the World. *Int. J. Intell. Syst. Appl.* **2017.**

Patch Management is the Catalyst for Growth in the IoT Industry, 13th Jun 2018, https:// datafloq.com/read/patch-management-catalyst-growth-internet-things/5106accessed on 17th feb 2019.

Pilkington, M. Blockchain Technology: Principles and Applications. *Research Handbook on Digital Transformations.* 2016, p. 225.

Roman, R.; Zhou, J.; Lopez, J. On the Features and Challenges of Security and Privacy in Distributed Internet of Things. *Comput. Netw.* **2013,** *57* (10), 2266–2279.

Sicari, S.; Rizzardi, A.; Grieco, L. A.; Coen-Porisini, A. Security, Privacy and Trust in Internet of Things: The Road Ahead. *Comput. Netw.* **2015,** *76,* 146–164.

Stanciu, A. Blockchain Based Distributed Control System for Edge Computing. In *21st International Conference on Control Systems and Computer Science Blockchain,* 2017; pp. 667–671.

CHAPTER 9

Attacks on Blockchain-Based Systems

TEJAS KHAJANCHEE[1*] and DEEPAK KSHIRSAGAR[2]

[1]*College of Engineering Pune (COEP), Pune, India*

[2]*Department of Computer Engineering and Information Technology, College of Engineering Pune (COEP), Pune, India*

Corresponding author. E-mail: khajancheet17.is@coep.ac.in

ABSTRACT

It was the 2008 Whitepaper penned by Satoshi Nakamoto and subsequently the January 3, 2009, that the world was introduced to the concept of decentralized cryptocurrencies through the Bitcoin's system. This cryptocurrency quickly gained a huge amount of popularity. As a result, many other cryptocurrencies were born following its success. The years that followed the rise of Bitcoin saw a huge involvement of research communities spanning individuals, academia as well as industry, experimenting and formulating sound theoretical foundations behind the concept. Later, Ethereum (Buterin, 2014) was introduced to the world in the year 2015. Ethereum went on to become the second most popular blockchain-based system because it practically exemplified that blockchain as a concept can be decoupled from the notion of cryptocurrencies through its capability of processing smart contracts.

After Bitcoin's birth, it didn't take long for the research community to recognize that cryptocurrency, even though it sounds very attractive, is not the main aspect of this system. The main aspect of this system was the Blockchain—a decentralized, distributed ledger that is maintained among a peer-to-peer network of computers through consensus. Research in the field of blockchain is sprawled out over a variety of subdomains—consensus, cryptography, proof-of-work, privacy, network—to name a few. In this chapter, the primary aim is to aggregate the existing attacks on blockchain-based systems. Consensus mechanism forms the spinal cord of any successful blockchain-based system. All other aspects of these systems have their roots somewhere in the consensus mechanism followed in that system. In earlier sections, this provides an intuitive foundation on the most pervasive consensus mechanism currently employed by the leading distributed-ledger-technologies in this world—the proof of work. In the later sections of the chapter, a historical account of the variety of attacks that have been performed against blockchain-based systems is covered. In the last section, the chapter highlights some of the proposed models of attacks that have not been demonstrated at large yet but, are in currently seen as a possibility and therefore, are under research.

This chapter will be aimed at affording the reader with an exposition catalog of the existing attacks on the blockchain-based systems, from a perspective of awareness.

9.1   INTRODUCTION

Nowadays, assuming a professional organization setting and getting a leave-application sanctioned is such an easy task. All you have to do is to go to your boss and get it done hoping that he/she will accept your reasons for taking the leave. What if the organization setting in which you are requesting leave is such that you must get the leave (more specifically, the "reasons" for taking the leave) validated/accepted by most managers present in your office for it to be officially considered as 'sanctioned'? Very obviously, in such a situation, the decision is to whether your leave will be sanctioned or not will depend on whether or not a greater proportion of managers "agree" upon accepting your reasons for availing the leave.

Intuitively, the notion of using a blockchain-based system for many different situations in our lives is similar to the one described above. This is a typical real-life example of a decentralized methodology or progress of operations. Blockchain-based systems have been demonstrated as being good models of realizing such decentralized progress of operations. Decentralization, therefore, becomes the central aspect of blockchain-based systems that exist today. More specifically, we are interested in knowing whether equally authoritative entities in a distributed, decentralized setting are in agreement with each other regarding a decision or the overall state of the system. It is therefore imperative that in order to understand and be aware of the kinds of security attacks that can be performed on blockchain-based systems, we first identify with decentralized systems and the aforementioned agreement (more commonly known in the blockchain's vocabulary as "Consensus") as the central aspects of a blockchain-based system.

9.2 A BRIEF HISTORY OF BLOCKCHAIN—SETTING UP THE STAGE

Blockchain was introduced to the whole world through the invention of a cryptocurrency known as Bitcoin (Nakamoto, 2008). The concept of blockchain became popular especially by that name because Bitcoin gained a massive amount of popularity, and "blockchain" happened to be the term used by Satoshi Nakamoto—the creator of Bitcoin—to describe the underlying technology over which Bitcoin was implemented. As a consequence of this popularity, there came a huge involvement of open-source tech-communities, industries, and individual enthusiasts. This led to heavy research, study, and reverse engineering of the matter and code available on the topic by Satoshi's implementation. A well-known outcome of this was that many alternative blockchain-based decentralized cryptocurrencies were born in the years to come, making this new technology more and more popular. Ethereum (Buterin, 2014), Litecoin (Litecoin, n.d.), Dogecoin (Dogecoin, n.d.), Namecoin (Namecoin, n.d., Kaldoner et al., 2015), Corda (Brown et al., 2016, Corda Official Documentation, n.d.) etc. are some of them to name a few.

An outcome of all these researches and studies parallel to all the popularity was a ubiquitous realization that blockchain is not one concept or a system or an algorithm. It is, in fact, a platform that is a very carefully architected permutation of a wide number of concepts, algorithms, and knowledge that together bring about the desired functionality of blockchain-based systems, the way we see them today.

9.3 CONSENSUS

As already established, consensus plays the most important role in the correct functioning of a decentralized distributed peer-to-peer system.

What are the mechanisms used by the modern blockchain-based systems to reach a consensus across the network? The most famous blockchain of the era—Bitcoin—and most of its derivatives use a novel algorithm called proof-of-work. It is important to remember that Bitcoin, and therefore block-chain is not some magical new technology. All of the constituent components that make up the blockchain system were had already existed approximately

15–20 years before Bitcoin was developed. It is just that nobody, till then had combined all these components in this particular manner to create such a system.

One of the most noteworthy things about the Bitcoin blockchain (and many other blockchains that followed) is the use of cryptographic hash functions, especially SHA256. This is a hashing algorithm that is capable of accepting an input of any size, mix it up completely and produce an output of a fixed size (see Fig. 9.1).

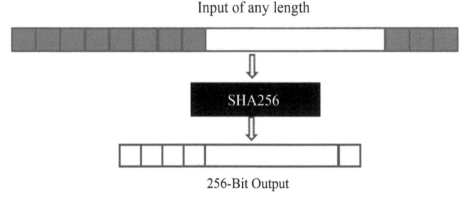

FIGURE 9.1   Trivial illustration of SHA256 fixed size output.

This output is unique to a certain data input. Therefore, a useful way of thinking about this is to consider this output as the fingerprint of the data pertaining to the famous analogy that for any person, his or her fingerprints are unique throughout the whole world. Also, there is one important feature about any good cryptographic algorithm that also persists in SHA256 and that is—you put any kind of data as input into it, you have no way of predicting what is going to come out of the algorithm. All you can be certain off is that it is going to be 256-bit long.

Moreover, when the output finally arrives, it looks absolutely random. No matter what statistical test you perform on the output of the SHA256 algorithm, it will always appear random. However, the output of this algorithm is deterministic. It means that, if you put the same data as input you will always get the same output. For example, given the input as "Hello World", the SHA265 will output the following:

String s = "Hello World"

#(s) = "a591a6d40bf420404a011733cfb7b190d62c65bf0bcda32b57b277d9ad9f146e"

The reader is encouraged to try this out and check. Note that input contains "H" and "W" in uppercase. Yes, the algorithm is case sensitive. Why? This is because the computer will interpret the letter as ASCII and therefore, uppercase and lowercase are inputs are different. Consequently, the bit-sequence going into the algorithm is also different. Just to make the point, given the above input again, all in lowercase, i.e. "hello world", we get:

String s = "hello world"

#(s)="b94d27b9934d3e08a52e52d7da7dabfac484efe37a5380ee9088f7ace2efcde9"

The point to ponder on here is that unique mapping of an input to an output of the SHA256. Due to this reason, this algorithm is used to verify the integrity of any kind of data. You can find the hash of a data you might have obtained from somewhere, verify it with the hash provided by the vendor of the data and be certain that the data obtained are indeed unaltered. Other algorithms also exist for such purposes such SHA1-family, MD5, etc.

There is a specific phrase mentioned in the text above—"any kind of data". Well what does that refer to? How will this algorithm apply to multimedia files (audio, video, image, etc.)? Such types of data are converted into text using something called base64 encoding or base32 encoding (Josefsson, 2003; 2006) and then supplied to SHA256. There is another important property that any strong cryptographic algorithm (Avalanche effect, n.d.) is expected to have—the avalanche effect. What this means is that there should be massive change in the output of a hash function even if there is a slightest change in the input (even a flip of a single bit). SHA256 does show this property. This is evident from the following example involving only a single bit flip. Let '#(x)' represent application of SHA256 algorithm on input data 'x'. Then,

#('2') = 'd4735e3a265e16eee03f59718b9b5d03019c07d8b6c51f90da3a666eec13ab35"

#('3') = '4e07408562bedb8b60ce05c1decfe3ad16b72230967de01f640b7e4729b49fce'

Here "2" in ASCII is 50 that in binary is 00110010 while "3" in ASCII is 51 that in binary is 00110011 that is a flip of a single bit. Clearly, it can be seen that the slightest change in the input to SHA256 gets the output to a totally different area in the 256-bit space, evidently proving avalanche effect in it.

So, realistic usage of this algorithm would correspond to few hundred thousand operations that do not take approximately more than half a second. The first innovative use of this algorithm was done in the year 1997, when Adam Back (Back, 2002) created a system for anti-spam called HashCash. The idea was to make any person who wants to send a message on the portal, calculate the hash of the message that the person is going to send before sending. This, as mentioned earlier, meant for the person to expend a few 100,000 operations spanning half a second before sending the message. Any other person viewing the message can simply verify by again doing those operations on their computer over the received message and be sure of a proof that the sender has legitimately performed this work before sending the message and that it cannot be a spam. This is okay if you are a legitimate user of the service, but if you are a spammer, you will probably want to send millions of messages in matter of seconds. Then, half second per message will very easily translate into half million seconds of work, which defies the logic of spamming. This was the first time a cryptographic algorithm was used as a proof-of-work mechanism.

What Satoshi Nakamoto (Nakamoto, 2008) did was that he made a little tweak in the way (Back, 2002) had applied proof-of-work and used it in Bitcoin making the task much more expensive and robust. He combined this mechanism to a peer-to-peer network (something that is very similar to how BitTorrent operates) to achieve consensus.

9.4   SATOSHI'S PROOF-OF-WORKS

So now we have it established that SHA256 is a strong cryptographic algorithm, to which we can pass any kind and amount of data and deterministically expect a 256-bit long output, which completely unpredictable and seemingly random. What is the probability that the first bit (MSB) of the output obtained is 0? It is 50%. We can carry forward this trend as follows. Let E_1 denote the event that *the first 'i' bits of the SHA256 output are zeros*. Then,

$$P(E_1) = 50\%; \ P(E_2) = 25\%; \ P(E_3) = 12.5\%$$
$$P(E_4) = 6.25\%; \ P(E_5) = 3.125\%; \ P(E_6) = 1.5625\%$$

This probability decreases exponentially as is evident in the following graph for a number of bits to be zero versus its probability.

E vs. P(E) Graph for Difficulty in Proof-of-Work

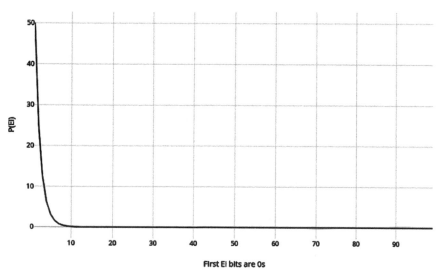

FIGURE 9.2 Graph of the number of bits from MSB set to be zero in the SHA256 output vs its probability.

Now since the output is a 256-bit long string, it can be interpreted as a number that going to be random. So essentially, the proof-of-work can be described as using SHA256 to produce a random number such that the number has a pattern to it. Let us consider the pattern to be that the first 16 bits are zero, or effectively the output is less than 2^{240}. We know that the output is deterministic but unpredictable. Keeping all those considerations, the only way this can be done is by adding some additional data (essentially a number) to actual input such that the output satisfies the constraints. Pertaining to the unpredictability of SHA256, the only way to do this is by checking the input along with different numbers one by one (brute-forcing) against the SHA256 and seeing after each run if the output is less than 2^{240}. This is the literal meaning of mining—*Finding a number which, when added along with the input and supplied to the SHA256 algorithm generates an output that is smaller than some target number.*

With regard to the graph above, this task becomes more and more diffi-cult (exponentially difficult) when the target number is smaller. In Bitcoin, this target number is reduced periodically thereby increasing the difficulty as time passes by. This is a typical game-theoretic design of the algorithm. What means is that there is no other way to find this number other than spending the required massive amount of energy to find this number (known as "Nonce" in the Bitcoin's vocabulary). It is a proof that if a node in the network is broadcasting this to its peers that it has mined a block, it has done at expense of the required amount of computing power. This forms an incentive for the peers to believe that the block is genuine thus reaching consensus throughout the network.

This was the one model of consensus mechanism that is used in most of the blockchain-based systems. Other models also exist. For example, Dash (Dash, n.d., Dash (crytocurrency), n.d.)—a cryptocurrency derived from Bitcoin, follows the Bitcoin's structure, but uses 'X11' as the Hash function instead of SHA256. This gives a very clear idea of how beautifully this concept was designed and how it involves so much of learning.

9.5   BAD REPUTATION DESPITE SUCCESS

It has already been established that blockchain technology has gained a massive amount of traction. The popularity of Bitcoin, Ethereum, and many other decentralized cryptocurrencies and platforms plainly signify that blockchain technology is on a firm path to success. Despite all this, there still exist two opposing forces–Devotees and the Security Research Community. Devotees are the chauvinistic practitioners, entrepreneurs, and enthusiasts who are consistently dispatching newer, competitive solutions to real world use-cases using the blockchain as a tool. And they have espoused this philosophy —*"Blockchains work in practice, but not in theory"*. They have accused the security research community (Bonneau et al., 2015) for failing to appreciate the technology for the novel invention it is, and still today shrugging it off just because it lacks meticulous theoretical groundwork. The research community now portrays to have reached a middle ground saying (Keenan, 2017)—*"The mathematical underpinning of all the prime blockchain-based systems is, probably if not provably, sound."*

Undeniably, even though blockchains have flourished positively over the world, they still have, unfortunately, maintained a bad reputation among a contending mass of people and governments across the world. Why is that? Some spectacular events have occurred over the decade of blockchain's rise that was scrupulously covered all over the media.

9.6   BAD CODE AND BAD PEOPLE

9.6.1   *BREAKDOWN OF MT. GOX*

A Cryptocurreny Exchange (also known as digital currency exchange) is a digital marketplace where traders can buy and sell digital currencies in exchange of other assets. These assets could also be other cryptocurrencies, or tangible assets, or it could be fiat currencies also. Mt. Gox—a Tokyo-based bitcoin exchange—and Mycoin–a Hong Kong-based bitcoin exchange failed arrestingly. The downfall of Mt. Gox afforded the development community with some of the most insightful what-not-to-do deterrents. Mt. Gox had lost a whopping $460 million worth cryptocurrency value in the year 2011 and completely died out of business. It was pointed out that that the internal software (Keenan, 2017) of the company was hideous blunder. There was absolutely no version control mechanism incorporated within the firm. According to an account, "if multiple developers happened to be working on the same file, anyone of them could accidentally have his/her colleague's code overwritten." It was as if the company was not at all aware of the concept of development sandboxes. Consequently, they were testing patches on the live production environment that handled actual money of their customers.

9.6.2   *BITSTAMP BITCOIN THEFT*

This is also a classic exemplar of the "people are your biggest problem" (Keenan, 2017) notion. Bitstamp is a chartered cryptocurrency exchange company based in Luxembourg, London. It offers trading, deposit and with-drawal services for cryptocurrencies as well as fiat currencies. It supports these services for USD, EUR, Bitcoin, Ethereum, Litecoin, and Bitcoin Cash. This company was founded (Bitstamp, n.d.) as a competitor to Mt. Gox in the year 2011. In 2012, the company was struck by a Distributed Denial-of-Service (DDoS) (Alisie, 2012) attack twice within a span of 12 h wherein the attacker had demanded a ransom of 75BTC. In 2014, the company was, yet again, struck by a DDoS attack (Pagliery, 2014; Kharif, 2014) this time forcing the company to suspend withdrawal services for a week. Later, in 2015, the firm suffered phishing attack directed at the company officials that resulted in the theft of almost 19,000BTC. The phishing (Whittaker, 2015) was carried via Skype calls. This evidently proves that blockchain-based systems are vulnerable to classic hacking techniques.

9.6.3   UPS AND DOWNS OF THE DAO

Let us first get an idea of what a Decentralized Autonomous Organization, in general, means. The word "governance" broadly expands to the process of making decisions and formulating rules that ultimately lead to the correct and desired functioning of an organization, society, or a system. The term is extremely broad and can be applied into many contexts. A decentralized autonomous organization (DAO)–also sometimes referred to as a decentralized autonomous corporation (DAC)—is basically an organization whose governance rules are delineated in the form of a computer program. Also, this program is kept transparent and is expected to be controlled by its shareholders, not influenced by the central government (Decentralized Autonomous Organization, n.d.). Some of the known DAOs are Dash (Dash, n.d.; Dash (crytocurrency), n.d.); Bitshares (Bitshares, n.d.; Schiessl, 2018); The DAO (The DAO (organization), n.d.); and Digix (Digix, n.d.). The DAOs that have existed till now have demonstrated that the mere existence of a DAO, its governance rules, and its financial transactions are maintained on a blockchain (Decentralized Autonomous Organization, n.d.). The legality of such a design of an organization (Decentralized Autonomous Organization, n.d.), (Popper, 2016a) is still debatable.

A Venture Capital is a type of financing that is afforded to small, early stage start-ups that are estimated to have, or have already demonstrated high growth potential. This financing is done by venture capital funds in trade of equity (Venture Capital, n.d.). The DAO was a decentralized autonomous organization launched as an investor-controlled venture capital fund. It was aimed at providing a new decentralized business model that would organize both commercial and non-profit enterprises. On April 30, 2016 (The DAO (organization), n.d.), it was launched on the Ethereum blockchain with no conventional management structure or board of directors. The DAO underwent a growth period in which anyone was allowed to send ether to a unique wallet address and get something called "DAO Tokens" in return (Falkon, 2017). During this time (The DAO (organization), n.d.), (Falkon, 2017), it gathered about a $150 million of funds—a feat that was marked as the world's largest crowdfunding ever.

It was not even a month of its launch that a paper was released that highlighted significant security pitfalls in the company's design (Popper, 2016b). Within less than 2 months from its launch, it was on June 17, 2016 that The DAO was hit. A crack was found by a hacker—the smart contract that was governing the functioning of The DAO would send the funds in ETH first and then update the internal token balance (Falkon, 2017). This vulnerability was also pointed out in the research paper that highlighted numerous other flaws (Popper, 2016b). The hacker deployed recursive calls to the smart contract effectively asking it to send ether multiple times before it could update the balance. This led in a drain of about $70 million worth ETH (The DAO (organization), n.d.), (Falkon, 2017). The whole drain was restored by the Ethereum's team by creating a hard fork of Ethereum's blockchain at the block 1920000 (The DAO (organization), n.d.; Buterin, 2016), but this move spoiled everything for The DAO since it was criticized for having the whole concept of blockchain to be undermined (Falkon, 2017). The learning here is that the bug was not in the Ethereum's code. The bug existed in the code of this application—The DAO—that was built on top of the platform.

9.6.4   THE ETHEREUM THEFT

In July 2017, "an unknown attacker exploited a critical flaw in the Parity multi-signature wallet on the Ethereum network" (Qureshi, 2017) ultimately stealing $31M US of the cryptocurrency. In response to this attack that was still on at that time, "a group of benevolent white-hat hackers from the Ethereum community rapidly organized" and drained the remaining $150M US that was at risk, on behalf of the rightful owners. It was an extraordinary event that equivalent to the good guys robbing all banks across the world to prevent a progressing robbery chain engulfing those banks too. The fault here was not with Ethereum or the Parity multi-signature wallet system–the cryptography was perfect. Rather, it was found that the default smart contract code that the Parity client gave to its new users had a serious vulnerability (Qureshi, 2017).

9.6.5   SILK ROAD AND ITS INVOLVEMENT WITH BITCOIN

This is taken up as a separate topic as it does not directly talk about an attack against the blockchain-based systems, in the sense it has been mentioned up until the previous section. Silk Road was an online shadow economy and the new age darknet marketplace, most popularly known for facilitating trade of illegal drugs (Silk Road marketplace, n.d.). The website used to operate using Tor services enabling the users to anonymously browse and participate in the trade. This website was launched in February 2011 and was shut down in October 2013 by the following the arrest of Ross Ulbricht (Silk Road (marketplace), n.d.)—the alleged founder and owner of the site who went under the pseudonym Dread Pirate Robert.

The analysis and revelations made following his arrest, uncovered the fact that all the transactions on this website were taking place in terms of bitcoins. Silk Road was designed in such a way that buyers and sellers used to be protected by the bitcoin's rise and fall (Greenberg, 2013; Tarbell, 2013). Even though during the time, bitcoin's pricing was very volatile, this did not affect the business that was running on Silk Road. This is because buyers and sellers were not interested in the worth of bitcoin, instead they were more fascinated by the fact that currency was not regulated by any central banks and does not require any online registrations, effectively affording a degree of anonymity to the participants (Greenberg, 2013).

According to a sealed complaint registered against Ulbricht, it was revealed that approximately $1.3 billion worth of cryptocurrency fluxed in the Silk Road network. Just imagine the size of the dark market that existed purely at the pride of Bitcoin's blockchains decentralized anonymity. A summary of the statistics is given in the following table.

TABLE 9.1 Statistics of Conspicuous Involvement of Bitcoin in Silk Road (Silk Road Marketplace, n.d.).

Detail description	Value	USD equivalent
Total transaction count	1,229,465 txns	N.A.
Total Bitcoin count	9,519,664 btc	$1.2 bn
Commission to Silk Road	614,305 btc	$79.8 mn
Total buyers	146,946 buyers	N.A.
Total vendors	3877 vendors	N.A.

The above historical accounts from the decade that followed blockchain's birth are not all, but some of the events worthy of being highlighted due to the kind of media attention they received. They clearly show that blockchain is not some magical new technology. It is a very humble beginning to a new kind of software paradigm, and like all other technologies do, blockchain is going through bumpy ride on a journey aimed at perfection. Blockchains currently are evidently vulnerable, and consequently there is a huge learning curve that exists for anyone who wants to jump into it. The following sections will cover an account of the potential weaknesses and some of the proposed attacks in the Bitcoin's blockchain protocol and some of the measures that have been incorporated in the newer protocol versions in order to mitigate these.

9.7 WEAKNESSES AND POTENTIAL ATTACKS IN BITCOIN BLOCKCHAIN

Till the previous section, we saw a real historical log of some of the most novel attacks that were performed on some of the popular blockchain-based systems and, also a side of the world of internet that exploited a mere feature of the Bitcoin cryptocurrency blockchain to their advantage in illegal trading. Let us now look more closely at some of the weaknesses that have been pointed in the Bitcoin's blockchain system, and some of the mitigation measures that the Bitcoin's protocol incorporates to avoid those to be exploited.

9.7.1 WALLET THEFT

It has already been demonstrated in (Keenan, 2017) and (Qureshi, 2017) that multi-signature wallets are prone to theft. In the bitcoin's system also, the wallet by default is stored in unencrypted format. Recent releases of the Bitcoin's client software have provided support for encryption (Ameya, 2018), but again, it dependent on the user whether he/she opts for it or not.

Also, it was noted in Ameya (2018) that an old wallet's along with its old password can be easily retrieved through the backup facility. The catch here is that even if you have updated your password, if your old account in the backup is accessed via your new password, and the attacker having all the funds siphoned from your account, this will also result in your current account as well. This will clearly make your account more vulnerable. The proposed solution to this was a change in the password-update process. When password is changed, a completely new wallet is created with new addresses and having all your funds transferred to this new wallet. So, even the backup account is compromised, your funds will be safe. But as a con, this will induce a lag in the password-changing process and will also leave your wallet without a backup (Ameya, 2018), for some initial time.

9.7.2   SYBIL ATTACK

In context of social media, this attack is like having one person creating multiple user accounts for malicious reasons. In this attack, the attacker will attempt to create multiple client nodes that are controlled by him. As a result you can be connected to a network that is controlled by the attacker without you even being aware about it (Ameya, 2018; Douceur, Druschel et al. 2002). The problem addressed by (Douceur et al., 2002) is that in the absence of a central authority, it would be troublesome for an honest node inside the peer-to-peer network to establish that it is connected to another honest node as his peer.

Extending to this, it will also be troublesome for the same honest node to even be sure whether all his peers are honest or there is a single remote node, presenting itself in the form of multiple identities, duping the honest node to think it is connected to different honest nodes. The possibilities after this could be as follows:

1.   The attacker can keep you away from being able to accept newer relayed transactions and blocks. This will lead you to be effectively disconnected from the network.
2.   You will be relayed the blocks that are only created by the attacker. This can make you prone to double-spending attacks.
3.   Theoretically, in an extreme case, this could lead to a 51% attack also wherein the attackers could change the order of the transactions, thus preventing them from being confirmed.

Blockchain-based applications use consensus algorithms like PoW that actually do not directly prevent the possibility of a Sybil attack, they just make it impractical to implement these attacks.

9.7.3   TIME-DRIFT ATTACK

This is also known as "time-jacking attack" (Boverman, 2011). If we consider a bitcoin block for example, it contains a list of transactions, the previous block's hash, the Merkle-root hash, the nonce of the mining, and along with some more information, it contains the timestamp representing the approximate time at which the block was created. The effect of timestamp is that it allows the system to regulate the supply (Boverman, 2011) of new bitcoins. Based on the timestamp of the blocks, a continuous monitoring of the time it took to create the previous block is kept, and pertaining to that, the difficulty level for the next block is adjusted so as to maintain a stable growth-rate of the blockchain at 1 block per 10 min.

How is this timestamp added knowing that the system is a peer-to-peer model? A counter that represents the network time (Boverman, 2011) is maintained by each node internally. This is based on the median time of the node's peers. However, the network sets to the system time automatically if the median time is found to be more 70 min apart from the system time.

This is based purely on a theoretical vulnerability observed in the Bitcoin's protocol. This kind of attack has never been demonstrated live on any blockchain. Each node can be thought of as generating two timestamp values one of which is finally accepted as the timestamp—T_n (network time) and T_s (system time). Bitcoin's system has some rules that are followed when handling timestamps. According to the acceptance rule of the timestamp at individual node-level, T_n and T_s are not allowed to be more than 70 min apart. If, for some reason, they do fall apart beyond the 70 min boundary, T_n is set to be equal to T_s. Let $T_{(s)}$ represents the final timestamp that is going to be given to the block. Mathematically,

$$|T_n - T_s| \leq 70\,min \tag{9.1}$$

$$T_{(s)} = \begin{cases} T_n, & |T_n - T_s| \leq 70\,min \\ T_s, & |T_n - T_s| > 70\,min \end{cases} \tag{9.2}$$

The attacker can create a Sybil network around its target having the majority as Sybil nodes. The attacker will then make the nodes pass inaccurate timestamps to the target node. If these timestamps are within range specified above, the target node's network time could be sped up or slowed down as per the attacker's whims and wishes. The following figures show the resulting scenario:

Panel (a) describes the normal functioning of the network. This is not the whole Bitcoin's network, but just a part of it. The remaining portion of the network is assumed to be working correctly. Here, all the peers are honest nodes ($H_n^{(i)}$). Therefore, they will not only pertain to the standard Bitcoin's protocol, but statistically, the network time ($T_n^{(i)}$—not depicted in the diagram above but, is implicitly assumed) values fired by them to each other will not be far apart from their counterparts at the destination nodes considering that the whole network has the knowledge of the previous mined block and its corresponding timestamp. The network therefore is in acceptance of a common/median timestamp value as the network time (T_n).

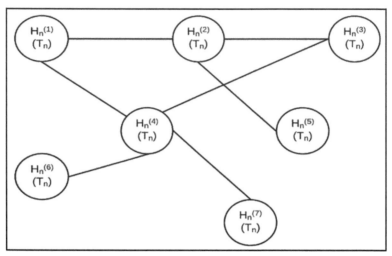

FIGURE 9.3A   Normal Scenario: All nodes are honest nodes ($H_n^{(i)}$) maintaining an acceptable network time as the actual time (T_n).

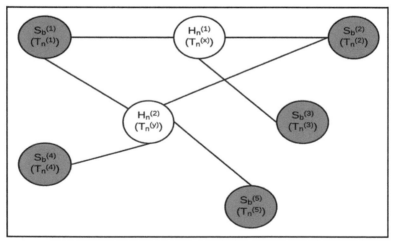

FIGURE 9.3B   Sybil Network Scenario conditioned for time-jacking. Five Sybil nodes ($S_b^{(i)}$) making a dominant presence over two honest nodes ($H_n^{(i)}$).

Panel (b) illustrates a typical Sybil scenario, where we have a part of the whole Bitcoin's network being Sybiled by the attacker whose is shown to be targeting two miner peers. The Sybil network could be created in such a way that there is no direct connection of the two honest nodes with each other. Here, the Sybil nodes are denoted as $S_b^{(i)}$, marked in red. They are shown to be generating inaccurate timestamps as $T_n^{(i)}$. Here, the following can be inferred:

1. The honest node $H_n^{(1)}$ is getting timejacked by the Sybil nodes $S_b^{(1)}$, $S_b^{(2)}$, $S_b^{(3)}$. This leads to the honest node developing a belief into an alternate time, thereby setting its own time to be different from the actual network-time as $T_n^{(x)}$.

2. Similarly, the honest node $H_n^{(2)}$ is getting timejacked by the Sybil nodes $S_b^{(1)}$, $S_b^{(2)}$, $S_b^{(4)}$, and $S_b^{(5)}$. This leads to this honest node also, developing a belief into an alternate time, setting its own to a different one as $T_n^{(y)}$.

The extreme case for the above scenario would be the ideal timejack achieved by the attacker. In this, one of the honest nodes could be made to slow down its clock exactly by 70 min while the other clock is sped up by 70 min; both individually satisfying acceptable time difference from their individual system times—$T_s^{(1)}$ and $T_s^{(2)}$, respectively. Assuming that $H_n^{(1)}$ was sped up and the $H_n^{(2)}$ slowed down, they could be actually made 140 min apart from each other. Mathematically,

$$T_n^{(x)} - T_s^{(1)} = 70\,min;$$
$$T_s^{(2)} - T_n^{(y)} = 70\,min;$$
$$T_n^{(x)} - T_n^{(y)} = 140\,min$$

Such disruption in the network can make victims prone to double-spending attacks, computational resource drains. This could also slow down the transaction confirmation rate. The victim nodes could be made to believe into another branch of the blockchain effectively disconnecting them from the honest network.

9.7.4 THE 51% ATTACK

This is a classic and the most talked-about attack in the blockchain space. This attack is mostly inspired by the existence of the bitcoin blockchain and its consensus algorithm—PoW. Due to the sheer network size of the bitcoin's system, this attack has not been demonstrated on it. This attack can be made possible when a single miner or a group of miners colluding together are able to take control over more than half of the network's hashing power (Frankenfield, 2019; Wong, 2018). If this happens to a PoW-based blockchain application, the attackers would then be able to:

1. Prevent transactions from gaining confirmation, effectively holding the authority in their hands whether to allow payments or not between users.
2. Revert transactions that were in confirmed earlier, making the double-spending possible.

This attack is one of the most feared attacks, in theory, for blockchain. Some of the cryptocurrencies based on blockchain technology and other blockchain-based systems typically based on PoW consensus have suffered from the 51% attack. These systems are relatively new with a very small market-footprint and digital-footprint. Krypton (Campbell, 2016) is a cryptocurrency that was launched in 2016 as a derivative from Ethereum in the form of a "sidechain." The same year it fell victim to a 51% attack after which it abandoned Ethereum and forked itself to become project of its own. Shift—another Ethereum-based blockchain—was struck by a 51% attack in 2016 (Redman, 2016). More recently, Bitcoin Gold (Wong, 2018)—a derivative of Bitcoin—and Verge (Joske, 2019) also suffered 51% attack in 2018 (Sedgwick, 2018). Verge was robbed off 250,000 verge (the unit of that cryptocurrency) incentivizing the technical team of verge to prepare a hard fork (Sedgwick, 2018).

It has been pointed out with these recent 51% hauls that blockchain-based systems, mostly the ones that are PoW-based and are very new in market, or are old but striving at a low market cap are more prone to 51% attacks since the hash-rates (Redman, 2016) are low due to lack of popularity.

9.8   CONCLUSION

This chapter covers many of the theoretically proposed attacks, and some of the many real-life attacks that have happened over blockchain-based systems in the decade of inception of this new technology. Enthusiasts in the field have got a lot to learn from events that have happened in the past. Some of these events have almost very strongly undermined the belief in this new technology. However, thorough analysis of those attacks and of the vulnerabilities they exploited has shown that in most cases, it was not the underlying concept that was insecure; it was a flaw in the design of the application that was built on top of the blockchain as a platform or a human error while implementing the concept. This can be correlated to the situation that we usually encounter in the use of internet—*If a website is badly designed, or is unresponsive, that does not mean the Internet is badly designed. All it means is that one website is not working as per our expectations.*

Blockchain is a novel new technology that afforded the modern world with a model of implementing distributed decentralized systems. It is still in its infancy just like the internet was in the 1990s. There is a lot of scope for research, improvement, and standardization for the field that is yet to happen. It is will take a highly excited level of involvement, trial-error and persistence to refine the technology to meet our needs due to lack of popularity.

KEYWORDS

- **Bitcoin**
- **blockchain**
- **decentralized systems**
- **distributed systems**
- **security attacks**
- **proof-of- work**
- **consensus**

REFERENCES

Alisie, M. Bitstamp under DDoS. *Bitcoin Magazine*, October 16, 2012. https://bitcoinmagazine.com/articles/bitstamp-under-ddos-1350360852/ (accessed February 10, 2019).

Ameya Sybil Attacks and the Byzantine Generals Problem [Blog post], July 21, 2012. https://medium.com/coinmonks/sybil-attack-and-byzantine-generals-problem-2b2366b7146b (accessed February 16, 2019).

Avalanche effect. (n.d.). In Wikipedia—the free encyclopaedia, https://en.wikipedia.org/wiki/Avalanche_effect (accessed February 16, 2019).

Back, A. Hashcash-A Denial of Service Counter-Measure [White paper], ftp://sunsite.icm.edu.pl/site/replay.old/programs/hashcash/hashcash.pdf (accessed February 10, 2019).

Bitshares. (n.d.). https://bitshares.org/ (accessed February 14, 2019).

Bitstamp. (n.d.). In Wikipedia—the free encyclopaedia, https://en.wikipedia.org/wiki/Bitstamp (accessed February 10, 2019).

Bonneau, J.; Miller, A.; Clark, J.; Narayanan, A.; Kroll, J. A.,; Felten, E. W SoK: Research Perspectives and Challenges for Bitcoin and Cryptocurrencies. *IEEE Symp. Security Privacy* **2015,** 104–121. http://ieeexplore.ieee.org/stamp/stamp.jsp?tp=&arnumber=7163 021&isnumber=7163005

Boverman, A. Timejacking & Bitcoin the Global Time Agreement Puzzle [Blog post], http://culubas.blogspot.com/2011/05/timejacking-bitcoin_802.html (accessed May 25, 2011).

Brown, R. G.; Carlyle, J.; Griggs, I.; Hearn, M. Corda: An Introduction [White paper], https://docs.corda.net/_static/corda-introductory-whitepaper.pdf (accessed February 10, 2019).

Buterin, V. A Next-Generation Smart Contract and Decentralized Application Platform, an introductory paper to Ethereum [White paper], https://github.com/ethereum/wiki/wiki/White-Paper (accessed February 10, 2019).

Buterin, V. Hard Fork Completed [Blog post], July 20, 2016. https://blog.ethereum.org/2016/07/20/hard-fork-completed/ (accessed February 14, 2019).

Campbell, R. Krypton Abandons Ethereum for Bitcoin Proof of Stake Blockchain after 51% Attack. *CryptoCurrency News,* May 09, 2016. https://www.ccn.com/krypton-ethereum-bitcoin-proof-of-stake-blockchain-after-51-attack (accessed February 20, 2019).

Corda Official Documentation. (n.d.). https://docs.corda.net/ (accessed February 10, 2019).

CrypoRect. Blackpaper Verge Currency 5th Edition [White paper], https://vergecurrency.com/static/blackpaper/verge-blackpaper-v5.0.pdf (February 20, 2019).

Dash (cryptocurrency). (n.d.). In Wikipedia – the free encyclopaedia, https://en.wikipedia.org/wiki/Dash_(cryptocurrency) (accessed February 14, 2019).

Dash. (n.d.). https://www.dash.org/ (accessed February 14, 2019).

Decentralized autonomous organization. (n.d.). In Wikipedia – the free encyclopaedia, https://en.wikipedia.org/wiki/Decentralized_autonomous_organization (accessed February 14, 2019).

Digix. (n.d.). https://digix.global/ (accessed February 14, 2019).

Dogecoin. (n.d.). https://dogecoin.com/ (accessed February 14, 2019).

Douceur, J. R.; Druschel, P.; Kaashoek, M. F.; Rowstron, A. I. T. The Sybil Attack. In *Revised Papers from the First International Workshop on Peer-to-Peer Systems* (IPTPS '01); Springer-Verlag: London, UK, 2002; pp 251–260.

Falkon, S. The Story of the DAO – Its History and Consequences [Bog post], December 24, 2017. https://medium.com/swlh/the-story-of-the-dao-its-history-and-consequences-71e6a8a551ee (accessed February 14, 2019).

Frankenfield, J. "51% Attack" in Investopedia, February 07, 2019. https://www.investopedia.com/terms/1/51-attack.asp (accessed February 20, 2019).

Greenberg, A. Founder of Drug Site Silk Road Says Bitcoin Booms and Busts Won't Kill his Black Market. forbes.com. https://www.forbes.com/sites/andygreenberg/2013/04/16/founder-of-drug-site-silk-road-says-bitcoin-booms-and-busts-wont-kill-his-black-market/#b7dfc816c423 (accessed February 14, 2019)

Josefsson, S. RFC4648: The Base16, Base32 and Base64 Data Encodings. The Internet Society, https://tools.ietf.org/html/rfc4648 (accessed February 16, 2019).

Kalodner, H.; Carlsten, M.; Ellenbogen, P.; Bonneau, J., Narayanan A. Citation: An Empirical Study of Namecoin and Lessons for Decentralized Namespace Design, *Workshop on Economics of Information Security.* http://citeseerx.ist.psu.edu/viewdoc/download?doi=10.1.1.698.4605&rep=rep1&type=pdf

Keenan, T. P. Alice in Blockchains: Surprising Security Pitfalls in PoW and PoS Blockchain Systems, *2017 15th Annual Conference on Privacy, Security and Trust (PST) 400-402*. http://ieeexplore.ieee.org/stamp/stamp.jsp?tp=&arnumber=8476964&isnumber=8476869

Kharif, O. Bitcoin Exhange Bitstamp Halts Customer Withdrawals. Bloomberg, February 12, 2014. https://www.bloomberg.com/news/articles/2014-02-11/bitcoin-exchange-bitstamp-halts-withdrawals

Litecoin. (n.d.). https://litecoin.org/ (accessed February 10, 2019).

Nakamoto, S. Bitcoin: A Peer-to-Peer Electronic Cash System [White paper], https://bitcoin.org/bitcoin.pdf (accessed February 10, 2019).

Namecoin. (n.d.). https://www.namecoin.org/ (accessed February 10, 2019).

Pagliery, J. Another Bitcoin exchange goes down, *CNN Business*, February 12 , 2014. . https://money.cnn.com/2014/02/11/technology/bitcoin-bitstamp/

Popper, N. A Venture Fund with Plenty of Virtual Capital, but No Capitalist, *The New York Times*, May 21, 2016. https://www.nytimes.com/2016/05/22/business/dealbook/crypto-ether-bitcoin-currency.html (accessed February 10, 2019).

Popper, N. Paper Points Up Flaws in Venture Fund Based on Virtual Money, *The New York Times*, May 27, 2016.https://www.nytimes.com/2016/05/28/business/dealbook/paper-points-up-flaws-in-venture-fund-based-on-virtual-money.html (accessed February 14, 2019).

Qureshi, H. A hacker stole $31M of Ether – how it happened, and what it means for Ethereum [Blog post], July 21, 2016. https://medium.freecodecamp.org/a-hacker-stole-31m-of-ether-how-it-happened-and-what-it-means-for-ethereum-9e5dc29e33ce (accessed February 16, 2019)

Redman, J. Small Ethereum Clones Getting Attacked by Mysterious '51 Crew', *news.bitcoin.com*, September 04, 2016. https://news.bitcoin.com/ethereum-clones-susceptible-51-attacks/ (accessed February 20, 2019).

Schiessl, S. The Bitshares Blockchain [White paper]. https://github.com/bitshares-foundation/bitshares.foundation/blob/master/download/articles/BitSharesBlockchain.pdf (accessed February 14, 2019)

Sedgwick, K. Verge Is Forced to Fork After Suffering a 51% Attack, *news.bitcoin.com*, April 05, 2018. https://news.bitcoin.com/verge-is-forced-to-fork-after-suffering-a-51-attack/ (accessed February 20, 2019)

Silk Road (marketplace). (n.d.). In Wikipedia – the free encyclopaedia, https://en.wikipedia.org/wiki/Silk_Road_(marketplace) (accessed February 14, 2019)

Tarbell, C. Sealed Complaint 13 MAG 2328: Unites States of America v. Ross William Ulbricht, September 27, 2013. https://web.archive.org/web/20140220003018/https://www.cs.columbia.edu/~smb/UlbrichtCriminalComplaint.pdf (accessed February 14, 2019).

The DAO (organization). (n.d.). In Wikipedia – the free encyclopaedia, from https://en.wikipedia.org/wiki/The_DAO_(organization) (accessed February 14, 2019).

Venture capital. (n.d.). In Wikipedia – the free encyclopaedia, https://en.wikipedia.org/wiki/Venture_capital (accessed February 14, 2019).

Whittaker, Z. Bitstamp exchange hacked, $5M worth of bitcoin stolen, *ZeroDayNet*, January 05, 2015. https://www.zdnet.com/article/bitstamp-bitcoin-exchange-suspended-amid-hack-concerns-heres-what-we-know/ (accessed February 10, 2019)

Wong, J. I. Every cryptocurrency's nightmare scenario is happening to Bitcoin Gold [Blog post], May 24, 2018. https://qz.com/1287701/bitcoin-golds-51-attack-is-every-cryptocurrencys-nightmare-scenario/ (accessed February 20, 2019)

CHAPTER 10

Blockchain and Bitcoin Security: Threats in Bitcoin

KAPIL CHOUHAN[1], PRAMOD SINGH RATHORE[1*], and POOJA DIXIT[2]

[1]ACERC, Ajmer, Rajasthan, India

[2]Dezyne Ecole College, Ajmer, Rajasthan, India

*Corresponding author. E-mail: pramodrathore88@gmail.com

ABSTRACT

We as a whole realize that blockchain is a procedure of bitcoin which is one of the most well-known technique for digital money that permit circulated add just open record all exchanges to be recorded. Appropriated agreement convention which depends on motivating force good evidence of work gives abnormal state of security this all framework and is controlled by a system hub called miners. Miners sincerely keep up the blockchain so as to get motivators. Bitcoin dispatch in 2009, its economy has developed at a tremendous rate, and it is presently worth about170 billions of dollars. This widespread rise in the market estimation of bitcoin incite enemies to achieve bad marks for benefit, and specialists to find new presentation in the framework, propose cure, and anticipate up and coming procedure. In this chapter, we present a dealt with examination that covers the security and assurance parts of bitcoin. Our strategy initially started from chart of Bitcoin Protocol, second critical portions of bitcoin, third parts convenience and last sections interchanges inside the system. The present chapter in bitcoin and its fundamental noteworthy progressions, for instance, blockchain-based agreement tradition lead to the execution of various security perils to the normal handiness of bitcoin. We discuss about the handiness and soundness of the front-line security courses of action. Moreover, we present current assurance and absence of definition examinations in bitcoin and analyze the security-related threats to bitcoin customers close by the examination of the present security shielding courses of action. Finally, we format the essential open troubles and propose rules for future research toward furnishing firm security and assurance frameworks for bitcoin.

10.1   OVERVIEW OF BITCOIN

Blockchain does not have a centralized record which is for safely trade computerized money, perform arrangements, and exchanges. Everyone in the system approaches the most recent same of encoded record so they can approve another exchange. Blockchain record is a gathering of all bitcoin replaced executed before. Essentially, it's a circulated database which keeps up a persistently developing sealed information structure squares which holds clumps of individual exchanges. The finished squares are included in a straight and sequential request. Each square contains a timestamp and data connect which focuses to a past square. Bitcoin is shared authentication in a system where every client associates with the enables each client to associate with the database and send new exchange to confirm and make new functions. This chapter clarifies the idea, qualities, need of blockchain and how bitcoin functions (Kanagavalli, 2017).

Bitcoin works over an inexactly associated P2P arrangement where hubs can join and leave the system. Bitcoin hubs are associated with the overlay organize over TCP/IP. At first, peers bootstrap the system by asking for friend address data from domain name system (DNS) seeds that give a rundown of current bitcoin hub IP addresses. Recently, associated hubs promote peer IP addresses by means of bitcoin address messages. Notice that a default full bitcoin customer sets up a limit of 125 TCP associations, of which up to 8 are active TCP associations. In Bitcoin, installments are performed by issuing exchanges that exchange bitcoin coins, alluded to as bitcoins (BTCs) in the continuation, from the payer to the payee. These substances are classified "peers," and are referenced in every exchange by methods for nom de plumes by bitcoin addresses. Each location maps to an exceptional open/private key pair; these keys are utilized to exchange the responsibility for among addresses. A bitcoin address is an identifier of 26–35 alphanumeric characters. Each bitcoin address is processed from an Elliptic Curve Digital Signature Algorithm (ECDSA) open key—for which the location proprietor knows the comparing private key—utilizing a change dependent on hash function. Since hashes are single direction function, it is conceivable to figure a location from an open key, yet it is infeasible to recover the open key exclusively from the bitcoin address. Review that, utilizing ECDSA marks, a friend can sign an exchange utilizing his or her private key; some other companion in the system can check the validness of this mark by confirming it utilizing the open key of the underwriter. A bitcoin exchange is shaped by carefully marking a hash of the past exchange where this coin was last gone through alongside the open key of things to come proprietor and fusing this mark in the coin. Exchanges take as info the reference to a yield of another exchange that spends similar currencies and yield the rundown of addresses that can gather the exchanged coins. An exchange yield must be recovered once, after which the yield is never again accessible to different exchanges. When prepared, the exchange is marked by the client and communicates in the P2P arrangement. Any friend can confirm the realness of a BTC by checking the chain of marks (Peck, 2017).

10.2 OBJECTIVES

Identifying the choices and members in an effective blockchain usage:

- Learn about different blockchain stages—open source and commercial
- Understand the facilitating and mining choices
- Awareness of related advancements
- Understand the essential programming languages, and
- Determining the choices encompassing the security of blockchains

10.3 SUPPORTED TRANSACTION TYPES

Bitcoin bolsters various default exchange types. Normally, just upheld exchange types are communicated and approved inside the system. Exchanges that don't coordinate the standard exchange type are for the most part disposed of. Note that since exchanges can have various yields, distinctive yield types can be consolidated inside a solitary exchange (Barski and Wilmer, 2013). Figure 10.1 shows single input spending.

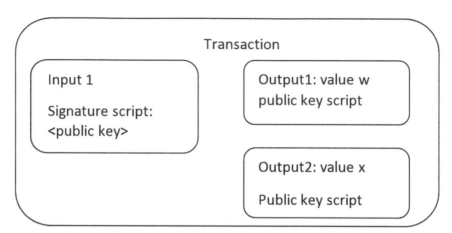

FIGURE 10.1   A transaction with a single input spending to two output addresses.

10.4   BUILDING BLOCKS AND CRYPTOGRAPHIC TOOLS

Here Bitcoin protocol has some protocols which are cryptographic tools to cryptographic hash functions such as SHA256 and RIPEMD160, Merkle trees, and the ECDSA.

10.4.1   CRYPTOGRAPHIC HASH FUNCTIONS

Hash function maps a subjective long info byte arrangement to a fixed size yield—regularly alluded to as a review—adequately fingerprinting the information grouping. Cryptographic hash function alluded to hash works that display two fundamental properties: onewayness, and impact obstruction. Let H: {0,1} *→ {0,1} n allude to a cryptographic hash work. Casually, the single direction property suggests that given H (x), it is (computationally) infeasible to determine x. Then again, the impact obstruction property suggests that it is computationally infeasible to discover x6 = y with the end goal that H(x) = H(y). The plot opposition property of cryptographic hash function comprises critical security columns in bitcoin. For instance, the PoW in bitcoin is for the most part dependent on processing hashes, and the id of an exchange compares to the hash of the exchange. Hash functions are a base segment of various sorts of information structures utilized in bitcoin (Khanaa et. al, 2013).

10.5 BITCOIN DATA TYPES

In this section, we introduce the main bitcoin specific data types:

10.5.1 SCRIPTS

Bitcoin presented a custom non-Turing complete scripting language trying to help distinctive kinds of exchanges and broaden the pertinence of exchanges past the straightforward exchange of assets. Contents are stack-based, bolster various function, and either assess to genuine or false. The language bolsters many diverse opcodes extending from basic examination opcodes to cryptographic hash function and mark check. Since contents should be executed on any bitcoin hub, they could be manhandled to lead disavowal of administration attacks and in this way, a significant number of opcodes have been incidentally handicapped. This was one of the principle reasons why contents don't give rich help when contrasted with standard programming languages (Khanaa and Thooyamani, 2014).

10.5.2 ADDRESSES

A location is an extraordinary identifier that plays the job of the originator or potentially the goal of some random exchange. A location's related bitcoin parity can be zero or positive, the greatest measure of bitcoins that can be made by the present accord convention. Regularly, bitcoin addresses begin with the decimal 1, or 3, and normally run from 26 to 35 characters.

10.5.3 TRANSACTIONS

An exchange is an information structure that can take at least one source of info and yields. Info reclaims the BTCs that are referenced in a previous exchange yield. Exchanges thusly successfully structure a chain of exchanges, and BTCs are actually just kept in exchanges yields, not inside addresses. An exchange yield determines what number of BTCs it contains just as under which conditions a resulting exchange can recover the yield. The ensuing exchange reclaims the yield of a previous exchange by encoding the important spending data in an exchange input. The conditions under which a yield can go through are encoded with the assistance of contents, and just the members that can give the right contribution to the content, to such an extent that it assesses to valid upon execution, are permitted to spend the BTCs yield by a given bitcoin exchange (Jeyalakshmi, 2017). Figure 10.2 shows transactional input.

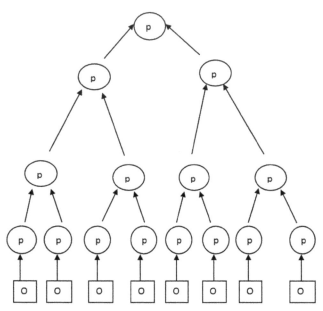

FIGURE 10.2   Transaction among inputs.

10.6   SECURITY CONCERNS AND RISKS RELATED TO BITCOIN

Bitcoin is so far the best cryptographic money. By and by, much the same as different digital forms of money, bitcoin has seen costs drop significantly for as long as couple of months. Value instability stays a standout amongst the most noteworthy difficulties confronting all cryptographic forms of money, as they attempt to explore a dubious biological community toward being perceived as world cash.

Virtual monetary standards have some security issues like money spending is not under control, and fears of rebel miners taking part in narrow minded mining. These worries that can be dangerous toward bitcoin are additionally genuine toward some other digital currency; however, not especially in a similar way (Kaliyamurthie et al., 2014).

Here's a concise once-over of a portion of these worries.

10.6.1   VULNERABLE WALLETS

There should be a major issue in the bitcoin wallets with respect to security attacks and robbery. There was a report from a group of scientists from Edinburgh University said they found frail spots in equipment wallets that can be misused. As pointed by a similar research, even the focused scrambled equipment wallets were as yet powerless because of that escape clause.

Utilizing malware, the researchers had the capacity to block correspondence between the wallet and PCs. This security break influences the protection of bitcoin clients on the grounds that their assets can undoubtedly be occupied to various records (Rebecca et al., 2015).

10.6.2   HACKERS AND CYBER ATTACKS

The potential for a devastating assault coordinated at bitcoin trades stays genuine. There have been noteworthy attacks on trades previously; yet, however, bitcoin's esteem drooped a short time later. Feelings of trepidation still proliferate of one that may totally handicap the prevalent cryptographic money. We are not discussing an assault on the blockchain itself; that is very nearly a nonstarter.

Bitcoin is likewise compromised by Distributed Denial of Service (DDoS) attacks. A report by Imperva showed that bitcoin exchanges had moved toward becoming top choices for DDoS attacks.

10.6.3   SELFISH MINING

Bitcoin's proceeded with utilization of confirmation of work accord system has another basic risk. With some mining pools ending up sufficiently unimaginable to direction noteworthy mining proportions, they will participate in slim minded mining.

Additionally known as square retention, a pool might utilize their process capability to mine a square and after conceal it from legitimate excavators as critical broadcasting the new square to the system.

The egotistical pool at that time endeavors to find the second square whereas the remainder grab in obscurity. Within the event that the ravenous mineworkers discern the way to find another square before alternate miners, at that time broadcasting the two squares makes the forked chain the longest.

The slim minded mineworkers are ahead of alternate miners, obtaining all of the rewards (Kaliyamurthie, 2014).

10.6.4   DOUBLE SPENDING

Despite the fact that fortifications have been initiated to relieve this serious concern, fears still flourish concerning this exchange hazard to bitcoin. Bitcoin is winding up progressively sturdier against facilitated double spends.

However, a few people may at present have the capacity to establish attacks that would make them profit by utilizing a similar coin twice in a similar exchange. For example, Bob buys things from Alice and sends Alice x bitcoins.

In the meantime, Bob executes a comparative exchange to a location he controls utilizing the equivalent bitcoins. In spite of the fact that Alice may trust that Bob has sent the cash and may not try to affirm, Bob's location might be credited with the exchange while Alice's won't get the pondered exchange (Khanaa, and Thooyamani, 2014).

Irreversibility at that point makes it trivial for Alice to get the exchange refuted. What's more, there is no plan of action in light of the fact that bitcoin is unregulated.

10.7   THE PROBLEMS WITH BITCOIN

As we push ahead in a world seeing new digital forms of money being made day by day, we need to be exceptionally clear. We fight that digital forms of money are digging in for the long haul; however, you ought to painstakingly think about which innovation is predominant before contributing. As most speculators know, bitcoin depends on blockchain innovation. However, there are a few reasons why bitcoin as of now isn't the best cryptographic money for worldwide applications.

To start with, there are numerous specialized entanglements with bitcoin, in spite of the fact that this contention applies to all cryptos. All things considered, contrasted with purchasing stocks or bonds (or even alternatives), digital money buys are not actually simple.

10.7.1   THE THREE MAJOR PROBLEMS WITH BITCOIN AND BLOCKCHAIN TECHNOLOGY

(a) **Scalability:** Bitcoin is interesting and inventive. A totally decentralized framework where money-related activities that require an abnormal state of trust can be executed without confiding in anyone. In any case, "trustless-ness" accompanies a cost: hard breaking points on the quantity of exchanges that can be prepared in a predefined time interim.

Because of the expanded appropriation of cryptographic forms of money as of late, the quantity of clients and exchanges have soar, testing the cutoff points of original blockchain frameworks.

Traditional blockchains like bitcoin are currently compelled in value-based throughput in light of the idea of their convention and blockchain plan. Right now, the essential plan structure of most existing frameworks is a straight connected rundown style blockchain. As selection increments and more excavators are pulled in to the framework, all mining ability is committed to mining the one next square in the straight blockchain. Along these lines, the expanded mining ability doesn't encourage adaptability by any means.

(b) **Limited programming ecosystem:** As an original, blockchains have set up a base programming environment. Blockchain-based smart contracts have additionally introduced another time of computational law whereby contracts are sponsored and conceded to by a blockchain which is impartial and generally common. Ethereum virtual machine was a huge advance up from the exceedingly constrained bitcoin programming condition. In any case, with expanding appropriation, the EVM has hit plan points of confine-ment and security entanglements. In this way, despite the fact that ethereum made the application improvement viewpoint on a blockchain, building complex applications stays extremely troublesome (Lamport, 2015).

(c) **Blockchain security vulnerabilities:** Credit should be given to original blockchains for their convention layer security. Not many occurrences of the bitcoin convention or the ethereum convention being helpless have been accounted for—which looks good for their security considering these systems have now been around for quite a long time.

Be that as it may, application layer security has been insufficient up to this point. Dull application layer security has brought about numerous occur-rences from the Parity Multi-Sig Wallet issue (which prompted 500,000 ethers being trapped) to the scandalous DAO assault—which caused an unavoid-able forking of ethereum into two distinctive blockchains and networks.

10.8   MULTIPLE APPLICATIONS OF BLOCKCHAIN

The conceivable outcomes present in blockchain innovation are the real explanation for financial specialists' and business visionaries' grip of bitcoin. Composing for the Huffington Post, a main tech official at Salesforce clarified that personality check dispatching and production network coordinations, fabricating, physician endorsed medication guideline, and a lot more procedures would all be able to be upgraded and better shielded through blockchain usage.

In what could fill in as a concise contention for the quest for blockchain improvement, Afshar stated, "blockchain capacity to digitize, decentralize, secure and boost the approval of exchanges."

Just like the case with such a significant number of new innovations or ideas, question in regards to its proficiency—just as solace and commonality with existing techniques, prompting watchfulness of progress—right now fills in as the fundamental driver for blockchain cynics.

As indicated by Quartz, parallels exist between the developing fervor around blockchain and the theory that powered the late 1990s/mid 2000s website blast. Knowing the past demonstrates to us how that didn't turn out well for a lion's share of its members, with just the most forceful organizations and dynamic items getting to be pillars of the innovation business. It could well be that hundreds or thousands of organizations and business people all attempt to create blockchain frameworks and most fall flat, leaving a chosen few at the leader of the pack—in spite of the fact that to some degree that is valid for some portions of the free market (McMillan, 2013).

10.8.1   ATTACK CASES

Here, we discuss about the attacks on blockchain and their systems, and then identify the various issues occur in these attacks.

10.8.1.1   SELFISH MINING ATTACK

The infantile mining ambush is driven by assailants (i.e., pompous diggers) to get undue rewards or misusing the preparing force of reasonable excavators. The attacker holds discovered squares furtively and after that tries to fork an individual chain. Sometime later, self important excavators would mine on this private chain, and attempt to keep up a more expanded personal branch than the open branch since they furtively hold even more recently discovered squares. In the meanwhile, reasonable excavators continue mining on the open chain. New squares mined by the attacker would be revealed when the open branch approaches the length of personal branch, with the ultimate objective that the real excavators end up wasting enlisting force and grabbing no reward, in light of the way that vain diggers circulate their new squares just before genuine excavators. Therefore, the biased diggers gain a high ground, and reasonable excavators would be supported to join the branch kept up by silly mineworkers. In the underlying situation of selfish-mine, the length of the open chain and private chain are the equivalent. The selfish-mine includes the accompanying three situations:

(1) The open chain is longer than the private chain. Since the processing intensity of childish miners might be not as much as that of the legit mineworkers, narrow minded excavators will refresh the 14 private chain as indicated by the open chain, and in this situation, egotistical mineworkers can't increase any reward.

(2) Fair mineworkers at the same time locate the main new square. In this situation, narrow-minded mineworkers will distribute the newfound square. Fair miners will mine in both of the two branches, while egotistical mineworkers will proceed to mine on the private chain. In the event that narrow-minded excavators right off the bat locate the second new square, they will distribute this square right away. Now, narrow-minded miners will increase two squares' rewards in the meantime. On the off chance that genuine mineworkers right off the bat locate the second new square and this square is kept in touch with the private chain, egotistical excavators will pick up the principal new square' rewards, and fair miners will pick up the second new square' rewards. Something else, if this square is kept in touch with the open square, legit mineworkers will pick up these two new squares' rewards, and narrow minded excavators won't increase any reward.

(3) After narrow-minded miners locate the main new square, they additionally locate the second new square. In this situation, narrow-minded miners will hold these two new squares secretly, and they proceed to mine new squares on the private chain. At the point when the primary new square is found by genuine mineworkers, narrow-minded excavators will distribute its very own first new square. At the point when fair mineworkers locate the second new square, the

narrow-minded miners will quickly distribute its own second new square. At that point narrow-minded excavators will pursue this reaction thusly, until the length of the open chain is just 1 more noteworthy then the personal chain, after which the egotistical miners will distribute its last new square before genuine mineworkers discover this square. Now, the private chain will be viewed as legitimate, and therefore egotistical excavators will pick up the rewards of every single new square.

10.8.1.2   DAO ATTACK

The DAO is a shrewd contract which executes a group financing stage. The DAO contract was assaulted simply after it has been conveyed for 20 days. The aggressor stole around 60 million US$. The assailant abused the reentrancy defenselessness for this situation. Right off the bat, the assailant distributes a malevolent brilliant contract, which incorporates a pullback work call to DAO in its callback work. The drawback will send to the called, which is moreover as call. Thus, the aggressor can take all the ether from DAO (Ciaian et al. 2016).

10.8.1.3   BGP HIJACKING ATTACK

Border Gateway Protocol (BGP) could be a true directive convention and manages, however, IP bundles square measure sent to their goal. To capture the system traffic of blockchain, attackers either influence or manage BGP directive. BGP capturing normally needs the management of system directors that may presumably be victimized to defer organize messages. It'll have a crucial impact. The assailants will with success half the bitcoin system, or put off the speed of sq. unfold. Assailants lead BGP seizing to catch bitcoin excavators' associations with a mining pool server. By rerouting traffic to a mining pool affected by the attacker, it had been conceivable to require cryptanalytic cash from the person in question. Since the BGP security expansions don't seem to be broadly speaking sent, prepare directors ought to depend upon observant frameworks, which might report maverick declarations. Be that because it might, despite whether or not associate degree assault is distinguished, subsidence a capturing still value hours because it could be a human-driven procedure comprising of modifying style or detaching the attacker.

10.8.1.4   ECLIPSE ATTACK

The eclipse attack enables an aggressor to hoard the majority of the unfortunate casualty's approaching and active associations, which detaches the injured individual from alternate friends in the system. At that point, the aggressor can channel the injured individual's perspective on the blockchain, or let the unfortunate casualty cost superfluous registering power on out of date perspectives on the blockchain. Moreover, the aggressor can use the injured individual's processing capacity to lead its very own malignant demonstrations. We consider two kinds of obscuration assault on Bitcoin's shared system, to be specific botnet assault and framework assault.

10.9   SECURITY ENHANCEMENTS

In this area, we outline security improvements to blockchain frameworks, which can be utilized in the advancement of blockchain frameworks. Figure 10.3 shows smartpool execution.

SMARTPOOL

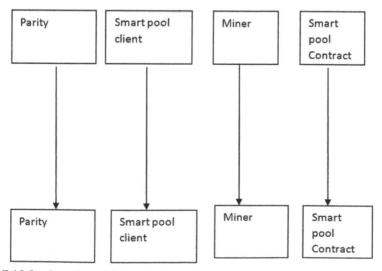

FIGURE 10.3   Overview of Smart Pool's execution process.

This represents a genuine risk to the decentralization nature, making blockchain helpless against a few sorts of attacks. Smart Pool gets the exchanges from ethereum hub customers (i.e., equality), which contain mining assignments data. At that point, the digger conducts hashing calculation dependent on the errands and returns the finished offers to the smart pool customer. At the point when the quantity of the finished offers compasses to a specific sum, they will be focused on smart pool contract, which is sent in ethereum. The smart pool contract will confirm the offers and convey rewards to the customer. Contrasted and the conventional P2P pool, smart pool framework has the accompanying preferences:

(1) **Decentralized**: The focal point of the Smart Pool is realized as canny contract, which is sent in blockchain. Excavators need at first interface with the client. Mining pool can rely upon client's agreement framework to run.

(2) **Adequacy**: Diggers can send the completed ideas to the smart pool contract in groups. Additionally, excavators simply need to send some segment of offers to be checked, not all offers.

(3) **Secure**: Smart Pool utilizes data structure, which can shield the attacker from resubmitting shares in different gatherings. Besides, the affirmation procedure for Smart Pool can guarantee that reasonable excavators will increment.

10.10   ENHANCEMENT OF INFORMATION TECHNOLOGY GENERAL CONTROLS

The executives of blockchain require explicit techniques to moderate the dangers identified with innovation that can be recognized, evaluated, checked, and controlled. While associations that receive blockchain will proceed to actualize and screen IT controls using institutionalized systems, key control improvements ought to be considered in the accompanying spaces:

1.   **Data security arrangements:** Policies for information the board, character the executives, and cryptography ought to be refreshed.
2.   **Physical security:** Blockchain will require proceeded with spotlight on equipment security and key administration.
3.   **Key administration and cryptography controls:** Organizations should concentrate on controls that help the age of cryptographic

keys; additionally, support, recharging, and security of the keys will be basic.

4. **PC activities:** Organizations will be required to invigorate their foundation security controls, including hostile to infection, zero-day abuses, back up and reestablish, and remote access.

5. **Coherent access controls:** Blockchain will require improvements of access controls like HR on-boarding/off—boarding and character-based access and key provisioning.

10.11   INTEGRATION OF BLOCKCHAIN INSTANCES

Execution of blockchain requires the innovation to be flawlessly coordinated with other blockchain arrangements, business frameworks, and advances. Thus, it is important that every interconnected application and frameworks speak with one another and interface information consistently. However, the general IT controls system, associations should concentrate on the accompanying regions when making arrangements for blockchain or executing/coordinating applications that depend on such innovation:

1. Data conversion and legacy system integration:

Blockchain arrangements inside an association in all likelihood will be coordinated with other inheritance stages like web servers, databases, centralized computers, and third party applications. Associations ought to investigate existing heritage applications, scrub and change inheritance information with the goal that it is clear by blockchain interfaces, and execute apparatuses and procedures to stack total and precise information.

2. Key administration for legitimate access:

Any client can get to just information that is put away inside their square and access to the square is administered utilizing open/private keys. This necessitates associations have successful key and advanced testament the executives foundation and arrangements set up to secure and keep up general society and private keys. Furthermore, associations considering open permission blockchains should oversee and ensure the uprightness of the accord calculation inside their individual surroundings. Associations that influence character and access the executives (IAM) arrangements inside their condition should decide how to adequately incorporate blockchain arrangements with their current IAM system and impression.

3. Access contemplations for equipment security:

Transacting blockchain information expects access to the record document or interfaces, which must be marked by a particular private key. These private keys are commonly put away on equipment-based tokens that are either connected to physical identifications, PIV/CIV cards, or biometric confirmation instruments. Security and the executives of the private keys require an all encompassing security approach that incorporates the different equipment instruments that help the open key foundation and physical security of the outside media used to store keys.

Blockchain is situated to change how associations work, oversee exchanges, and structure and sell items and administrations. At last, the innovation can empower organizations to bridle the intensity of man-made consciousness to complete monetary action (Buterin, 2013).

10.12   SECURITY

In this area, we talk security risks what's greater security implications of the Bitcoin Protocol what's more structure arrangement. Since bitcoin might be propelled specific money for remarkable business esteem, motives on abuse shortcomings to profit need help all inclusive.

Cryptography and wallets: So as to use bitcoin, the Initially Relic a customer needs might be a wallet. The wallet holds an open/private key pair, which might be the best close estimation of the client's record. Subsequently, clearly, it is indispensable to degrade defensive means should verify the wallet.

Double spending: Coins that need help unmistakable (e.g., Eventually Tom's examining sequential numbers), double. Utilizing concerning delineation discussed before will be inconsequentially recognizable. Monetary forms consistently additionally recognized logged off circumstances (Vega, 2013), which irritated it boundless on contact the bank to Commission those exchanges. This made double contributing a main problem, paying little mind to a fundamental power existed. For the most part, the bitcoin setting will be a web circumstance.

Pooled mining: Transaction adaptability originally, exchange adaptability similarly as pointed out previously, bitcoin must endeavor will care for a structure on which no substance controls the greater part of the computational vitality. Essentially of the whole deal in bitcoin's history, this might have been the situation. Be that as it may, being the foremost will viably check a piece happens best. For a little probability thusly, those outcome to solo mining will be extraordinarily blasted: a basic reward yet just uncommonly. Some of the time miners along these lines a consistently expanding sum gets together under mining pools. On a mining pool, different miners help them. Square period conjointly. Every part seeks portions of the nonce space for a considerable nonce. In the occasion, a champion among them will be fruitful, the advantage might be conferred. Consequently, every part gets constant little rewards, instead of sometimes, gigantic ones. There are a couple of decisions which handle those issues. As a matter of fact, it might be in the personality or premium about every excavator will remain with that scattered organic network soundness. In this manner, the most effortless outcome is that miners. In the end Tom's scrutinizing themselves switch should diverse mining pools something like that likewise will reasonably redistribute those power. Correspondingly, as it turned out before, this meets desires amazingly well: calls eventually Tom's scrutinizing the gathering keeping should switch mining. Composes those scattered frameworks for partners to specific, we will consider those bitcoin convention, the coming about exchange models besides their implications for larger part of the information expansion. Moreover, we social affairs give a point of view investigating bitcoin enlivened sort out. Arrangements furthermore benefits, for instance, elective space name and illuminating structures. Those bitcoin framework need expects which balance beginning with the people of peer-to-peer record conferring structures.

10.13   VULNERABILITIES

Double spending: Double Spending is a conceivable method to cheat in an exchange. A client endeavors two exchanges all the while by exchanging an amount of bitcoins to one individual and exchanging the equivalent definite bitcoins to another. This parts the square chain exchanges must be confirmed multiple times over and the right square chain will win the split. The current answer for double spending assault is square chain, which tells the client what pulling forces check and can be trusted. In any case, a superior answer for keeping this assault is to have a timestamp server.

Brute force: Brute Force Attack is only a thorough key hunt. In a hypothetical way, a savage power assault can be utilized against any information, which has been encoded. This sort of attack can be utilized when every single other choice have been utilized to abuse the shortcomings of the encryption framework.

Finney attack: A Finney assault works just if the shipper who is associated with the exchanging acknowledges unverified exchanges. It can likewise work when the merchant trusts that couple of moments will guarantee that everybody in the system concurs that he has to be sure paid additionally the Finney assault can possibly happen when the assailant is mining and is keeping control of the substance of the square. The programmer who is utilizing this system makes two exchanges one recognizing the individual he is attempting to rip off and one crediting himself. This takes a specific measure of time and the minute he is fruitful, he utilizes the put away first exchange to buy the great he needed. At that point he liberates the premised square. This guarantees the previous exchange is refuted, however, when it appears in the square table. The best thing about this assault is that it is for all intents and purposes imperceptible until the whole procedure has been executed and it is past the point of no return for the merchant. The main arrangement that is accessible for this sort of assault is to make no less than one affirmation for each exchange before conveying the products and rehashing this procedure for every single exchange (Buterin, 2013)

10.14   FUTURE OF THE BLOCKCHAIN TECHNOLOGY:

Blockchain innovation will in all respects likely effect organizations crosswise over ventures, including budgetary administrations, social insurance, oil and gas, retail, stimulation, promoting, media, vitality, and the open division. In monetary administrations, for instance, blockchain can be utilized to make worldwide installments, and exchange stocks, bonds, and wares. It can likewise give a review trail to controllers. Blockchain can likewise be utilized to make new types of advantages and to exchange existing illiquid ones, for example, versatile minutes, vitality credits, and preferred customer credits.

It is generally trusted that blockchain will be a problematic innovation that will empower monetary administrations, medicinal services, government, and different businesses to offer trust as an administration. All the more vitally, blockchain has introduced the period of a "programmable economy" at the end of the day, a worldwide market controlled by algorithmic organizations and associations that keep running on blockchain-based systems and use rules encoded in delicate product or man-made reasoning to take part in financial movement.

10.15 CONCLUSION

In this chapter, we have described the working standards of distributed cryptographic monetary forms and particularly security and we have appeared in spite of the fact that the cryptography behind bitcoin isn't as of now broken, the framework can be assaulted with a great deal of PC power or malignant growth hubs. These attacks are troublesome and, in all actuality, programmers follow bitcoin clients to take their wallets with malevolent programming. We likewise demonstrated that bitcoin isn't intended to be unknown, yet a client who needs to keep his/her private personality can basically do it. For bitcoin upgrades and extra alleviations, we give thoughts to hub evaluating clients in the system not to keep customers from the believed exchange branch database produced by the attackers. We offer an approach to ease conceivable issues brought about by attacks with a ton of registering power by logging squares and adding checkpoints to the squares chain. Future work will be to incorporate further research on relief measures and their usage, just as a numerical verification of bitcoin's cryptographic security.

KEYWORDS

- **bitcoins**
- **blockchain**
- **cryptocurrency**
- **security threats**
- **user privacy**

REFERENCES

Barski, C.; Wilmer, C. Bitcoin for the Befuddled, USA: No Starch Press, 2013.

Blockchain.info/spoofed transactions problem/Aug. Reddit, 2015. [Online]. https://www.reddit.com/r/Bitcoin/comments/3fv42j/blockchaininfospoofed_transactions_problem_aug_4/ (accessed Jan 01, 2017).

Buterin, V. OzCoin Hacked, Stolen Funds Seized and Returned by StrongCoin—Bitcoin Magazine, Bitcoin Magazine, 2013 [Online]. https://bitcoinmagazine.com/articles/ozcoin-hacked-stolen-funds-seized-and-returned-by-strongcoin-1366822516/ (accessed Dec 30, 2016).

Ciaian, P.; Rajcaniova, M.; Kancs, D. The Digital Agenda of Virtual Currencies: Can Bitcoin Become a Global Currency? *Inf. Syst. e-Bus. Manag.* **2016,** *14,* (4), 883–919.

Chambers, J.; Yan, A. W.; Garhwal; Kankanhalli, M. Currency Security and Forensics: a Survey. *Multimedia Tools Appl.* **2014,** *74,* (11), pp. 4013-4043.

Fischer, M. et al. Impossibility of Distributed Consensus with One Faulty Process. *J. ACM* **1985,** *32,* (2), 374–382.

Garay, J. A. et al. Bootstrapping the Blockchain—Directly, Cryptologyeprint Archive, 2016 [Online].

Hals, T. Mt. Gox files U.S. Bankruptcy, Opponents Call it a Ruse, Reuters, 2014 [Online]. http://www.reuters.com/article/us-bitcoin-mtgox-bankruptcy-idUSBREA290WU20140310 (accessed Dec 30, 2016).

https://eprint.iacr.org/2016/991.pdf (accessed Dec 30, 2016).

http://en.wikipedia.org/wiki/Bitcoin#wallets (accessed Dec 6, 2014).

http://visual.ly/bitcoin-infographic (accessed May 24, 2012).

http://www.coindesk.com/information/comparinglitecoin-bitcoin/ (accessed Apr 2, 2014).

https://www.weusecoins.com/en/miningguide (accessed 2014].

Jeyalakshmi, G.; Arulselvi, S. Remote Procedure Calls in Access Points. *Int. J. Pure Appl. Math.* **2017,** *116,* (15) Special Issue, 523–526.

Jeyanthi Rebecca, L.; Anbuselvi, S.; Sharmila, S.; Medok, P.; Sarkar, D. Effect of Marine Waste on Plant Growth. *Der Pharmacia Lettre* **2015,** *7,* (10), 299–301.

Kaliyamurthie, K. P.; Parameswari, D.; Udayakumar, R. Malicious Packet Loss During Routing Misbehavior-Identification. *Middle East J. Sci. Res.* **2014,** *20,* (11), 1413–1416.

Kanagavalli, G.; Sangeetha, M. Intelligent Traffic Light System for Reduced Fuel Consumption. *Int. J. Pure Appl. Math.* **2017,** *116,* (15) Special Issue, 491–494.

Kanniga, E.; Selvaramarathnam, K.; Sundararajan, M. Kandigital Bike Operating System. *Middle East J. Sci. Res.* **2014,** *20,* (6), 685–688.

Karthik, B.; Selvaraj, A. Noise Removal Using Mixtures of Projected Gaussian Scale Mixtures. *Middle East J. Sci. Res.* **2014,** 20, (12), 2335–2340.

Karthik, B.; Selvaraj, A. Test Data Compression Architecture for Low Power VLSI Testing. *Middle East J. Sci. Res.* **2014,** 20, (12), 2331–2334.

Karthikeyan, R.; Michael, G.; Kumaravel, A. A Housing Selection Method for Design, Implementation & Evaluation for Web Based Recommended Systems. *Int. J. Pure Appl. Math.* **2017,** *116,* (8) Special Issue, 23–27.

Khanaa, V.; Thooyamani, K. P.; Udayakumar, R. Elliptic Curve Cryptography using in Multicast Network. *World Appl. Sci. J.* **2014,** *29,* (14), 264–269.

Khanaa, V.; Thooyamani, K. P.; Udayakumar, R. Impact of Route Stability under Random Based Mobility Model in MANET. *World Appl. Sci. J.* **2014,** *29,* (14), 274–278.

Khanaa, V.; Thooyamani, K. P.; Udayakumar, R. Modelling Cloud Storage. *World Appl. Sci. J.* **2014,** *29,* (14), 190–194.

Khanaa, V.; Thooyamani, K. P.; Udayakumar, R. Patient Monitoring in Gene Ontology With Words Computing Using SOM. *World Appl. Sci. J.* **2014,** *29,* (14), 195–199.

Kleineberg, K.; Helbing, D. A Social Bitcoin Could Sustain a Democratic Digital World, 2016.

Lamport, L. The Part-Time Parliament. *ACM Trans. Comput. Syst.* **1998,** *16,* (2), 133–169.

Lamport, L.; Massa, M. Cheap Paxos, in International Conference on Dependable Systems and Networks. *IEEE Comput. Soc.* **2004,** 307.

Lamport, L. et al. The Byzantine Generals Problem. *ACM Trans. Program. Lang. Syst.* **1982,** *4,* (3), 382–401.

McMillan, R.; Metz, C. The Rise and Fall of the World's Largest Bitcoin Exchange, Wired. com, 2013 [Online]. https://www.wired.com/2013/11/mtgox/ (accessed Dec 30, 2016).

Nakamoto, S. Bitcoin: A Peer-to-Peer Electronic Cash System. https://bitcoin.org/bitcoin.pdf.

Peck, M. Bitcoin: The Cryptoanarchists' Answer to Cash. Retrieved April, 2014. http://spectrum.ieee.org/computing/software/bitcoin-thecryptoanarchists-answer-to-cash.

Some Miners Generating Invalid Blocks, Bitcoin.Org, 2015 [Online]. https://bitcoin.org/en/alert/2015-07-04-spv-mining (accessed Jan 02, 2017).

Sompolinsky, Y. et al. SPECTRE: A Fast and Scalable Cryptocurrency Protocol. Cryptology eprint archive, 2016 [Online]. https://eprint.iacr.org/2016/1159.pdf (accessed Dec 30, 2016).

Steadman, I. Wary of Bitcoin a guide to Some Other Cryptocurrencies%22. Retrieved April, 2014 http://arstechnica.com/business/2013/05/wary-ofBitcoin-a-guide-tosome-other.

The Great Chain of Being Sure About Things, Economist.com, 2015 [Online]. http://www.economist.com/news/briefing/21677228-technology-behind-bitcoin-lets-people-who-do-not-know-or-trust-each-other-build-dependable (accessed Dec 30, 2016).

Understanding the bitcoinj security model, Bitcoinj.github.io. [Online]. https://bitcoinj.github.io/security-model#pending-transactions (accessed Jan 01, 2017).

Vega, D. http://heavy.com/tech/2013/12/bitcoin-vs-litecoinpeercoin-ripple-namecoin/ (accessed Dec 02, 2013).

CHAPTER 11

Security and Privacy Issues in Blockchain Technology

D. PREETHI, NEELU KHARE*, and B. K. TRIPATHY

School of Information Technology and Engineering, VIT University, Vellore, Tamil Nadu, India

Corresponding author. E-mail: neelu.khare@vit.ac.in

ABSTRACT

Blockchain technology has a plethora of applications ranging from cryptocurrency to today's smart contracts. Blockchain has eight important elements, such as decentralization, consensus model, transparency, open-source, identity and data security, autonomy, immutability, and anonymity. Security and privacy are vital aspects of Blockchain technology. Security in blockchain is dealt with by preserving the transactional data in a block against internal, peripheral, and unintentional threats. Security can be ensured by detection and protection of threats, providing an appropriate response to the threats by applying security tools, policies, and IT services. Privacy in blockchain involves the ability to perform transactions without leakage of identification information. Although there are a number of studies which focus on the security and privacy issues, there lacks a brief examination on the security of blockchain systems. In this chapter, we present an elaborate study on the security and privacy to the blockchain. Furthermore, this chapter provides the challenges and the recent advances are briefly discussed and also suggest some future research directions in this area.

11.1 INTRODUCTION

Nowadays blockchain invites attention in both industrial applications and academic research. As one of the most familiar cryptocurrencies, Bitcoin has succeeded with its capital market cap of 67,999,590,426 dollars in 2019 (https://Coinmarketcap.com) with a particular structure for storing data, the transactions in Bitcoin are possible without hindrance from the third party. The origin to build Bitcoin is the blockchain technology, which is proposed by Sakamoto in 2008 and implemented in 2009 (Nakamoto, 2008). Blockchain perhaps characterized by the distributed public ledger and all the transactions are brought together as a list of blocks. These blockchain build-up as new blocks continuously added to it. For attaining the security to its user and the ledger organization in the blockchain, distributed consensus algorithms and the asymmetric cryptography are applied. The blockchain technology generally has key elements, such as decentralization, consensus model, autonomy, immutability, anonymity, transparency, open source and identity, and access (Joshi. et.al, 2018). With such characteristics, the blockchain can be extremely economical and achieve better performance. Blockchain is applied in various fields including finance (Swan, 2015; Fanning and Centers, 2016; Fahmy, 2018; Casey et al., 2018), healthcare (Alhadhrami, 2017; Siyal, 2019), internet of things (IoT) (Rahulamathavan, 2017; Banerjee, 2018), education, automobile industry (Fraga-Lamas and Fernandez-Carames, 2019; Dorri et al.), and defense.

Even though blockchain has its promising development in today's Internet technology; it is prone to several security and privacy challenges. This chapter's objective is to introduce this new topic that gains interest as well as to bring up to date information on blockchain. This study also includes security and privacy issues and challenges in the blockchain technology. Furthermore, a detailed description of various blockchain-based applications is presented in this chapter. First, scalability, which is the primary issue, needs to be addressed. The block size in bitcoin is restricted to 1 MB, where each block for about every 10 min is mined. Eventually, it is incapable of dealing with the high frequency trading applications in which the Bitcoin network is limited to a transaction speed of 7 numbers of transactions per second. Although, the larger blocks refer to an increase in storage capacity and with slow network propagation. This in turn results in the centralization and gradually with a small number of users who wish to maintain such a large blockchain. Thus, there exists a trade-off between security and block size and the security has been a tough challenge in the blockchain usage. Through selfish mining, the blockchain miners can gain income compared with their invested amount. Furthermore, it has been seen that leakage in privacy may also arise in the blockchain. Even with users' transactions being performed with their public and private keys. Moreover, current consensus algorithms, like proof of stake, proof of work, are also having some significant problems.

We structured this chapter into the following five sections. Section 11.2 will present the issues and challenges in blockchain technology. Section 11.3 deals with an in-depth description of the security and privacy of blockchain technology. Section 11.4 describes various blockchain-based applications and provides security and privacy issues for the same. Finally, Section 11.5 presents the conclusion and future directions in blockchain technology.

11.2   ISSUES AND CHALLENGES IN BLOCKCHAIN

This section describes various issues and challenges in blockchain technology in detail. Figure 11.1 depicts issues and challenges in Blockchain technology.

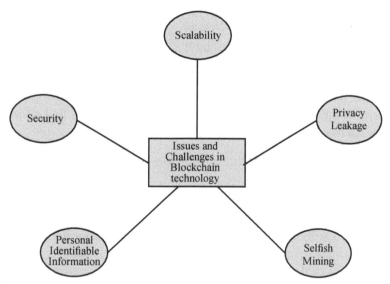

FIGURE 11.1   Issues and challenges in blockchain technology.

11.2.1   SCALABILITY

With the rapid increase in the number of transactions every day, the blockchain turns heavier. Bitcoin blockchain particularly performs transactions with a speed of seven transactions per second. Limitations on block size and the time taken to build a new block make the blockchain incapable to process the millions of transactions. Due to very small block size, the miners go for the transactions with high payment and, thereby, the transactions with low payments get delayed. But blocks with large block size make the blockchain propagate with a slower speed which results in the development of branches in the blockchain. Thus, makes the scalability difficult in blockchain. There are several attempts performed to confront the scalability issue in the blockchain, they are optimizing storage capacity in blockchain and redesigning the blockchain.

11.2.2   PRIVACY LEAKAGE

Privacy perhaps described as the ability to share information with its intended groups by an individual. Blockchain users utilize an alias, address to process their business transactions. The main consequence is that the blockchain transactions may reveal the user's information and they are user's way of using data, purchasing habits, and frequently visited places. The blockchain privacy is tied with anonymity and untying of user's transactions. The anonymity in blockchain can be achieved by not associating a user to a specific transaction. On account of this, every transaction in the blockchain is associated with an altered address. Also, untying the user's transactions can be achieved by dealing with the blockchain transactions and addresses, which are not associated with the real user identity.

11.3   SELFISH MINING

The blockchains are vulnerable to attacks from intriguing selfish miners. An approach called selfish-mine was proposed against attacks which compels the genuine miners to execute on the old public blocks branch.

Usually, it is defined that nodes more than 51% of computing power can flip side the blockchain and its transaction history. But, from the recent study, it can be seen that nodes with less than 51% computing power make blockchain insecure. In the study by Eyal and Sirer (2014), it can be seen that the blockchain becomes weak when a small amount of hashing power is exploited. In this selfish mining approach, selfish miners preserve information about their blocks which are being mined by not broadcasting to anyone and only on the fulfillment of some conditions, the information about the private branch can only be exposed to the public. Due to the longer private branch than the existing public branch, it could be given access to all its miners. Earlier to private blockchain broadcast, all the genuine miners are being exhausted with their resources by utilizing unnecessary branch, whereas these selfish miners without opponents are able to mine their private chain. Thus, these selfish miners' gains more profit. The rational miners could be drawn to these selfish pools and, thus, the selfish miners can obtain the computing power 51% easily.

11.3.1 PERSONALLY IDENTIFIABLE INFORMATION

Blockchain technology can be used for securing personally identifiable information without the need for a trusted centralized authority. It is provided by creating an identity on the blockchain for individuals with proper access control over their personal information. By utilizing the decentralized blockchain, a digital ID can be created with identity information. Every online transaction of an individual is associated with a digital watermark called digital ID. These IDs are used by the organizations to verify its their identification during each transaction, thereby reducing the fraudulent rate. With blockchain, users can utilize authentication apps than classic methods of using user id and password. This app can store the encrypted identification information which can be shared among companies and use on their own terms and conditions.

11.3.2 SECURITY

Security and privacy play an important role in any type of information system. Secure systems can be achieved with the collaboration of availability, confidentiality, and integrity. Security can also be obtained by combining authorization, authentication, and identification. Majority attack (51% attacks) and fork problems are some of the security problems in blockchain technology (Lin and Liao, 2017). Majority attacks occur when someone achieves a computation power of 51%, get access to the nonce value before others and gain control on the block. This may lead to security issues and it is termed as majority attack (51%) attack. The fork issues are associated with the version of a decentralized node when the agreement is given during upgradation of the software. During the new release of the software, the consensus rules in the newer agreement are also changed with respect to the nodes. Two types of forks in blockchain are hard fork and soft fork. The hard fork refers to a system with a newer version or agreement and which is not suitable with its earlier version, which leads to disagreement of older nodes with the newer nodes and result in two different chains. Soft fork refers to a system with a newer version or agreement and which is not suitable with its earlier version; the newer nodes disagree with the mining of older nodes.

11.4   SECURITY AND PRIVACY ISSUES IN BLOCKCHAIN TECHNOLOGY

In this section, we discuss the security and privacy in detail with the blockchain technology.

11.4.1   SECURITY IN BLOCKCHAIN

Blockchain security is described as the protection of transactional data in a block against the internal and peripheral, malevolent, and unintentional threats. Subsequently, this can be achieved with threat detection, threat prevention, and response to the threat by utilizing the services from the IT industry, security tools, and policies. In the context of blockchain, integrity can be obtained by providing the guarantee to its users with immutable transactions and, generally, integrity can be verified using the cryptography techniques. The blockchain achieves availability by granting its users to create communications with many users and their blocks are maintained using a decentralization mechanism which has several copies of the blockchain in the network.

Next, to ensure confidentiality in the blockchain, pseudoanonymization technique is used where the hash functions are used as a duplicate identity of the user. The blockchain structure ensures these three: authentication, authorization, and auditing functions, in which it allows users with their private keys can perform their transactions and all such transactions are allowed for audit purposes and made publicly available. The nonrepudiation provides assurance during its user's actions such as money transfer, message transfer, and purchase authorization. Also, the user cannot deny his actions where all the transactions are digitally signed.

11.4.2   PRIVACY IN BLOCKCHAIN

The privacy in blockchain can be strengthened by building it highly challenging from using others cryptographic information. A countless number of discrepancies can be obtained during the application of blockchain technology. Some of the common characteristics are described as follows (Zyskind and Nathan, 2015):

➢ Sorting data to be stored:
   Blockchains are flexible in storing various forms of data. On the perspective of blockchain privacy, it differs between personal and organizational information. However, rules for privacy are suitable for personal information, very tough privacy rules are applied for sensitive organizational information.

➢ Append only characteristic:
   In the undetected blockchain, it is difficult to change the contents of the earlier blocks. The append-only characteristics of blockchain in certain situations are not suitable when recording data incorrectly. Special concentration should be given when allowing the rights to the data subject to blockchain technology.

➢ Distributed Storage:
   The full nodes store full copies of the nodes in the network of the blockchain. Data redundancy occurs with these full nodes with the combination of the one of the blockchain characteristic. This data redundancy supports two main attributes of blockchain technology such as transparency and verifiability. The level of transparency and verifiability for an application in a network can be obtained by the data minimization techniques.

➢ Private blocks vs public blocks in the blockchain:
   The access to blockchain is extraordinary from the privacy point of view. In the advanced level, the nonpermit data on a block perhaps encrypted by authorized users upon conditional access where every node has a stored copy of the complete blockchain.

➢ Nonpermissioned vs permissioned types of blockchain:
   Applications which are public or nonpermissioned type of block-chain, all the users are allowed to store data. The permission to the trusted mediators has a direct effect on the distribution of control over the network.

11.5   SECURITY AND PRIVACY CHALLENGES AND SOLUTIONS FOR VARIOUS BLOCKCHAIN APPLICATIONS

In this section, we study the various blockchain technology-based applications. Also, we have a detailed discussion on some of the security and privacy challenges in various blockchain-based applications. Figure 11.2 shows the various blockchain applications mainly used.

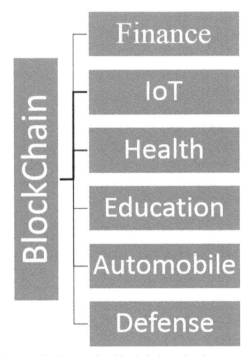

FIGURE 11.2   Various applications using blockchain technology.

11.5.1 *BLOCKCHAIN IN FINANCE*

Blockchain technology plays an important role in optimizing the global financial infrastructure and building an efficient economic system. The latest announcement from analysts of KPMG has described the blockchain benefits in the finance sector, they have higher efficiency, decrease in the rate of fraudulent and loss, enhanced customer satisfaction, and higher capital availability. The primary motivation for blockchain businesses is to combine both traditional banking and financial markets with cryptocurrency. Figure 11.3 depicts the application of blockchain technology in finance. Ripple Labs focuses on this newer technology to provide improvements in the current banking system and allows the traditional financial organizations to precede their business effectively. Ripple allows the bank in transferring the funds and also provides direct foreign exchange transactions without an intermediary. Other commercial systems are also incorporating the usage of bitcoin to traditional finance and market solutions for payments. An example is PayPal as it has been developed in parallel with Bitcoin. PayPal was at first developed as a new payment market solution different than that of the traditional financial market services, like Bitcoin. However, it has changed into a more regular business within the governed industry, where it is involved in the collection and verification of personal details of its customers. PayPal stood on the edge of its innovation in the financial sector; on the later, it started to focus like a corporate sector and, thereby, lost its credibility with respect to is opponent bitcoin (Swan, 2015).

The blockchain was first applied in the current financial system for handling the back-office transactions. These back-end activities are so costly to the financial institutions and, thereby, the financial services needed a technological breakthrough to perform these activities. Blockchain can be the breakthrough which can solve these financial transactions (Fanning and Centers, 2016).

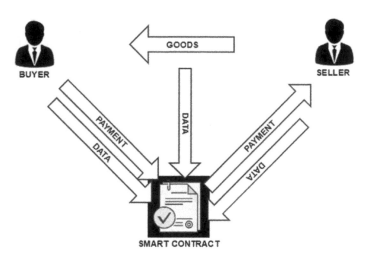

FIGURE 11.3   Application of blockchain technology in finance.

Furthermore, the smart contracts are introduced to the blockchain, where the financial contracts are automated on the blockchain. Financial contracts known as derivatives are particularly well suited for blockchain implementation. By automating the financial derivatives, it is possible to improve efficiency, increased visibility of market operation for regulatory organizations, and reduce transaction costs (Fahmy, 2018). Subsequently,

as in payments and foreign money returns is also one of the main signifi-
cances of blockchain-based applications to draw its focus. Blockchain-based
protocols used are called Ripple, an interledger protocol, to provide real time
cross border payments by combining the existing bank ledgers. It is far clear
that at least in finance, the production scale will be better as permissionless
blockchain introduced by Satoshi in 2008. All the initiatives of blockchain
are within the financial areas of applications are noted from outside of both
cryptocurrency and decentralized crypto exchanges, which are built with
small numbered nodes chosen by the system initiators. In some situations,
only one node is allowed to perform the validation of its network.

Permissioned blockchain mitigates challenges of scalability, privacy, and
governance, which are dealt by the public blockchain. The computational
capacities are not required much in securing the network, since validators are
incentivized. There is no need to contest for hash computing in cryptocurrency, despite it normally done with using their shared interest to obtain its
security. For regulatory reasons, identification of these actors is very much
significant. The involvement of several attendants exists, as the permissioned
ledgers are being relying on one or more trusted party. Despite the development of a single ledger-shared infrastructure in which every attendant has
access to these remedies and can remove reconciliation of multiple ledgers
and the activities are streamlined. But still, privacy, security, and scalability
concerns are to be addressed, where the processes involved in business and
the regulations undergo an examination for the overall financial system to
acquire a permissionless setup. Thereby, the permissioned blockchain acts
as the default choice for financial functionaries. Financial regulators need to
understand the implications of such models for financial the market structure
and develop the necessary rules to secure against unstable finance, reduce
fraudulent activities, promote competition, and encourage innovation.
Blockchain technology has its advantages both in upgrading centralized
systems and making it available for many users (Casey, 2018).

11.5.2 BLOCKCHAIN IN IOT

The latest IoT systems mainly rely on centralized and brokered models of
communication. IoT privacy concerns are complex and, also, it has insufficient standards for authorization and authentication of various IoT devices.
Standards involved in the promotion of security configurations comprise
virtual IoT platforms and, thereby, provides immature multitenancy. The
requirements involved in the new security technologies is necessary for

protecting the IoT platforms and devices from both information-based attacks and physical tampering, to encrypt their communications and to observe the security challenges like impersonating of "things" or denial-of-sleep attacks that drain the batteries, which leads to denial-of-service attacks (DoS). A primary example for the need of newer technologies for security is the latest distributed denial of service attack (DDoS) that infected the servers of top services like PayPal, Twitter, NY Times, and Netflix across the U.S. dated on 21st October 2016. It was occurred as a result of a gigantic thrash which is engaged by internet addresses in millions and malevolent softwares. These typical scenarios heightened fears of cybersecurity and an increase in internet security breaks. All evidence suggest that innumerable IoT systems, such as closed-circuit cameras and smart-home devices, were attacked by the malicious software and are used to exploit its servers.

A decentralized approach allows IoT issues especially in networking are described above. Any of the decentralized approaches should allow the following three basic functions:

➢ Allows peer-to-peer communication
➢ Sharing of files is distributed in nature
➢ Coordination of autonomous systems

In spite of all its advantages, the blockchain model has the following challenges:

➢ **Scalability**: Scalability issues appropriate to the blockchain results in centralization, which are developing a veil on the future of cryptocurrency.
➢ **Processing power and time**: Necessary to carry out encryption involved in a blockchain-based ecosystem.
➢ **Storage**: Blockchain avoids the necessity for a centralized server for storing the transactions and digital IDs of each device, whereas the ledger needs to be stored on their nodes. This leads to an increase in ledger size with time passing around.
➢ **Lack of skills**: A few people understand the working of blockchain technology and when IoT is added, eventually the number of skilled people will decrease.
➢ **Legal and compliance-related issues**: There is no compliance or legal code to observe, which challenges manufacturers and service providers in using blockchain technology.

On the development of the IoT solutions, it requires unique collaboration, connectivity, and coordination in the IoT ecosystem. All of the devices should be integrated with all other devices and work together, and all the devices should be able to interact and communicate with connected infrastructures and systems. But these are very time consuming, complex, and very expensive. A lightweight scalable blockchain (LSB) is proposed by Siyal et al. (2019). The authors have explored LSB in a smart home as an example for IoT-based applications. The LSB obtained decentralization from the formation of network overlay, in which it promotes end-to-end security and privacy by managing a public blockchain working with large resource devices. LSB utilizes many techniques for optimization algorithms for the purpose of lightweight consensus with distributed trust and to manage throughput.

Although blockchain provides fairly integrity and nonrepudiation, despite failed to preserve the confidentiality and privacy of the data or the devices involved. To address issues respect to privacy in blockchain, newer privacy in blockchain preserving architecture for IoT applications is based on attribute-based encryption (ABE) techniques that are proposed by Rahulamathavan (2017). In the study by Ahmad Khan and Salah (2018), significant IoT-based security issues are discussed and also showed blockchain technology plays a vital role in rectifying several IoT-based security issues. The authors also identified some challenges and open research issues based on security in IoT such as limited resource utilization, wide-ranging devices, security protocols with interoperability, single points of failures, vulnerabilities on hardware or firmware, updated management and trust, and vulnerabilities on blockchain.

11.5.3   BLOCKCHAIN IN HEALTHCARE

The blockchain technology today makes rampant strides in expanding its utility in various areas of applications, including the biomedical field. The blockchain is renovating the traditional healthcare practices with a much reliable healthcare system in the criteria of diagnosis and treatment in an effective way by sharing patient data in a secure and safe manner. In the upcoming years, the blockchain technology perhaps helps in the secure, authentic, and a personalized healthcare system by linking the complete clinical profile about a patient's health in real time and providing it with an up-to-date secured healthcare model. Figure 11.4 depicts the Blockchain transactions in healthcare.

The most common applications of blockchain technology in healthcare are currently in the area of electronic medical records, clinical research, medical fraud detection, neuroscience research, and pharmaceutical and industry research. Electronic health record (EHR) typically stores highly-sensitive and critical data related to patient's health, which is often frequently shared among clinicians, radiologists, healthcare providers, pharmacists, and researchers, for effective diagnosis and treatment. This highly-sensitive patient information during storage, transmission, and distribution among several entities, the patient's treatment can be compromised, which can imply severe threats to patient health and in maintaining the up-to-date patient history. A range of issues, including data privacy, data integrity, data sharing, record keeping, patient enrolling, and so on may arise in clinical trials. Blockchain, being the next internet generation, can provide effective remedies to these problems. Medicinal drug supply chains are vulnerable and consist of holes for fraudulent attacks. Blockchains provide a safe and secure platform to eliminate this problem and, in some cases, prevent fraud occurrence, by introducing higher data transparency and improved product traceability. Hyper ledger, a research foundation, launched a counterfeit medicine project using blockchain technology as a foremost tool, for inspecting and fighting the production of counterfeit drugs. Neuroscience research blockchain, as an innovation, brings several upcoming applications incorporating brain augmentation, the re-enactment of the brain, and brain thinking. Digitizing a whole human brain clearly requires some medium in which to store it, and it's here that blockchain innovation raises its head.

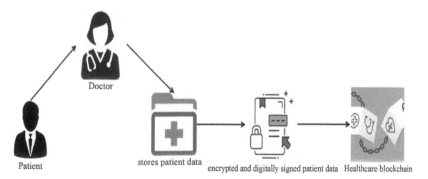

FIGURE 11.4   Blockchain transactions in healthcare.

In the study by Xia et al. (2017) proposed a model where blockchain helps patient to maintain, share, and have control over their overall health data with easy and secure privacy-preserving technique. This model is based on applications which are dealt with a secure multiparty computing (MPC) and indicator-centric schema (ICS). MedRec is a project initiated between the MIT media lab and Beth Israel Deaconess medical center. This platform supports a decentralized approach in providing authorization, permission, and sharing data among healthcare systems. The MedRec platform has been validated as a proof of concept with medical data and the developers are focusing on enhancing the project's extent by the addition of various types of data, users, and data contributors.

Healthcare is fully dependent on the patient's information for further treatment and issues on the security and privacy of data emerge continuously with this data processing. Privacy in healthcare involves the processing of individual patient's data in private or there arises the need authorization for accessing the data. Security in healthcare perhaps related to the process of securing sensitive data from intruders and eavesdroppers. In the action of preserving the healthcare data, the authenticated person gets access to the system for storing and retrieving the data. The communication should be done in a secure manner among the patient and the system. In this interaction, due to lack of security, there is a possibility of losing patients' confidential data, as there are a number of intruders available in the network to hijack all these secure data. In some cases, the loss of healthcare data has proven very dangerous. With the latest emerging attacks on healthcare systems, different countries are facing the destructive loss of data. Such attacks might loot the sensitive patient's health data easily as these are nonencrypted data that are stored on the server. A cyber attacker sometimes enters into the data preserving system and changes the personal patient data into insecure. Accountability, integrity, pseudonymity, security, and privacy of healthcare data should be controlled by the systems. Nowadays, patients are losing interest in EHR systems as it is vulnerable to privacy and security. Thus, the focus on the accountability and integrity of the EHR systems should also be considered. Pseudonymity of the patient is very crucial, as it contains sensitive personal healthcare data. The traceability and controllability play a vital role in the privacy-preserving systems.

11.5.4   BLOCKCHAIN IN EDUCATION

Blockchain promotes a highly prominent standard for the educational field. The most representative example is learning as earning (LAE) is an initiative developed by the institute for the future and the ACT Foundation.

The main goal of this initiative is involved with the Edublocks, which are the same as that of credit hours which are utilized nowadays for assessing and recording the student's learning ability. Edublocks not only used for student's ability tracking but also useful in assessing and recording informal learning as well, which includes the diversity and inclusiveness of training, competitions, students events, presentations on research, experience in an internship, and community service, etc., Edublocks constructed as a chain of distributed ledger or e-Portfolio, exclusively for the sake of students to gain credits for their learning anywhere, anytime, on-campus or off-campus. The collection of all Edublocks comprises of a student's credits achieved during their study are converted as an e-portfolio at their graduation comprising of all the earned income recorded from their various skills. The student can utilize this e-portfolio as a reference for their employers to hire them or in their interviews during their job hunt. In a learning as earning institution, the academic instructors will concentrate on helping the students to achieve most incomes in their e-portfolios. The advantage of this learning ledger includes the course assessment, improvements in the curriculum, and scholarship for financial assistance. It is significant to see that the complete LAE idea is developed using the blockchain technology. Blockchains allows a platform to improve transparency and trust among various stakeholders from higher education, including faculty, students, employers, administrative staff and alumni. As e-portfolios emerge from the data repositories using robust tools for improving student learning, the next generation student assessment tool, measuring the quality of an educational program, and recording the outcomes of the student's learning ability, which are developed using Blockchain technology. The future works can be the development in the areas of state of the art and applications in Blockchain technology and also can be applied to the applications involving higher education.

11.5.5 BLOCKCHAIN IN DEFENSE

Currently, many proof-of-concept works are underway to implement blockchain technology for the aerospace and defense industries. The primary defense-centric application of blockchain technology is cybersecurity. However, blockchain involves a distributed database, with no single point of failure or attack. Any altered instance of the database will provide information regarding a hacker's target, making it easier to track and combat the attacker. Moreover, sensitive information would be protected through the public/private key cryptography scheme at the heart of any blockchain implementation. Asset management is another potential blockchain application for defense department supply chains.

To account for the classified nature of some supply chains, a completely private blockchain could require invitations for a user/node to join the network. The blockchain could also be a hybrid system where select information is visible publicly but classified information is reserved for private invitees. The infrastructure itself could be handled by servers located at stateside military installations. Through this implementation, a complete cradle-to-grave audit trail, verifiable by all network users, could be generated to justify budgets and increase public confidence in the defense department's asset management process. In addition, current blockchain developers are touting potential improvements in verifying regulatory compliance, asset tracking, counterfeit protection, and recall management for defense supply chains.

11.5.6   *BLOCKCHAIN IN AUTOMOBILE INDUSTRY*

This section describes how the automotive industry has greatly transformed society, with development evolving from hybrid cars, electric cars, and self-driving smart cars which are additions to IoT applications. Despite the complex nature of the automobile industry, it requires the involvement of several industries driven 4.0 technologies, such as advanced manufacturing systems, robotics, cyber-physical systems, and augmented reality. One of the recent technologies that can take advantage of the automotive industry is blockchain, which can improve the data security, privacy, authentication, anonymity, accountability, traceability, transparency, integrity, robustness, and trustworthiness. It can provide sustainability for the long term and higher efficiency in industrial operations. The importance of applying the blockchain technologies to the automotive industry, ensuring its cybersecurity attributes have been reviewed and analyzed in the study by Fraga-Lames and Fernandez-Carames (2019).

Various existing applications of blockchain in the automotive industry are a remote software update, insurance, smart charging services, electric-based vehicles, and car-sharing services (Dorri et. al.,). The upgradations of the electronic control units (ECU) of a vehicle or fixing a bug installed in the software on one of the ECU's are termed as the wireless remote software update (WRSU). Securing these WRSU is one of the most crucial challenges in the ecosystem of auto motives, as it needs complete access to the vehicle and its embedded control systems. Insurance companies have started to offer flexible insurance fees to vehicles for their responsible customers. The insurance company has the identification information of the owner's vehicle which is stored in the cloud storage. This threatens the privacy of the owner's vehicle, as the data exchanged may involve sensitive data with respect to privacy, for example, vehicle location. The innumerable numbers of electric vehicles are rapidly built and this trend increases the demand for efficient and fast charging infrastructure for such vehicles. The interconnection of these smart vehicles to its owner's smart home devices and mobiles can provide many useful services.

The sharing car service, say for example Car Next Door, is rapidly developing which provides highly distributed services and it may require the interconnecting the smart vehicles, sharing car service providers, and the users of the services in a trusted and secure way. A trusted communication channel is necessary to exchange secure data including the vehicle location, keys for unlocking the car, and payment details of the user. The inherent properties of blockchain technology are ideally suited to reducing business friction and increasing trust among organizations across the automotive supply chain. Collaboration and communication among the participants in the supply chain are much improved through shared processes and record keeping. In the near-term, the technology is ready to be deployed by manufacturers and suppliers to protect their brands against counterfeit products and also to achieve significant cost savings through supply chain process improvements.

11.6   CONCLUSIONS AND FUTURE DIRECTIONS

Blockchain has shown its potential in industry and academic research with its key elements: Decentralization, consensus algorithms, transparency, open-source, identity and access, anonymity, and audibility. In this chapter, we present a comprehensive overview of security and privacy in the blockchain. We listed some recent issues and challenges in blockchain technology. Furthermore, we included some challenges and issues that would impede blockchain development. We also narrated some existing approaches for tiding over the stated humps. A new era dawns with blockchain-based applications springing up in various fields such as finance, IoT, healthcare, automobile, defense, education, and so on. We conducted some investigations on blockchain-based applications in these different areas. The future scope in blockchain technology should focus on the following areas: blockchain testing stops the tendency to centralization, big data analytics, artificial intelligence, and blockchain application. Although, there still exist some limitations in blockchain as discussed in this chapter and many innovative applications are difficult to be implemented. Blockchain will usher in, the due popularity to become the technology that everybody would move towards as it mellows. We have shown how blockchain technology is applied in umpteen applications on the grounds of security and privacy in this chapter.

KEYWORDS

- **blockchain**
- **privacy**
- **security issues**
- **challenges**

REFERENCES

Angraal, S.; Krumholz, H. M.; Schulz, W. L. Blockchain Technology Applications in Healthcare. *Circ. Cardiovasc. Qual. Outcomes* 2017. DOI: 10.1161/CIRCOUTCOMES.117.003800.

Alhadhrami, Z.; Alghfeli, S.; Alghfeli, M.; Abedlla, J. A.; Shuaib, K. Introducing Blockchains for Healthcare. *International Conference on Electrical and Computing Technologies and Applications (ICECTA)*, 2017.

Banerjee, M.; Lee, J.; Choo, K.-K. R. A Blockchain Future for Internet of Things Security: A Position Paper. *Digi. Commun. Net.* 2018, 4, 149–160.

Casey, M.; Crane, J.; Gensler, G.; Johnson, S.; Narula, N. The Impact of Blockchain Technology on Finance: A Catalyst for Change. Geneva Reports on the World Economy 21, 2018.

Dorri, A.; Steger, M.; Kanhere, S. S.; Jurdak, R. Blockchain: A Distributed Solution to Automotive Security and Privacy. 2017, *IEEE Commun. Mag.* 55 (12), 119–125.

Dorri, A.; Kanhere, S. S.; Jurdak, R.; Gauravaram, P.; LSB: A Lightweight Scalable BlockChain for IoT Security and Privacy. 2017, pp. 1–17.

Eyal, I; Sirer; E. G Majority is not enough: Bitcoin mining is vulnerable. In Proceedings of Financial Cryptography, Barbados, 2014.

Fahmy, S. F. Blockchain and its Uses, 2018, 1–10.

Fanning, K.; Centers, D. P. Blockchain and Its Coming Impact on Financial Services. *J. Corp. Account. Fin.* 2016, 27 (5), 53–57.

Fraga-Lamas, P.; Fernández-Caramés, T. M. A Review on Blockchain Technologies for an Advanced and Cyber-Resilient Automotive Industry. Special Section on Advanced Software and Data Engineering for Secure Societies. *IEEE Access*, 2019, pp. 17578–17598.

Joshi, A. P.; Han, M.; Wang, Y. A Survey on Security and Privacy Issues of Blockchain Technology. *Math. Found. Comput.* 2018, *1* (2), 121–147.

Khan, M. A.; Salah, K. IoT Security: Review, Blockchain Solutions, and Open Challenges. *Fut. Gen. Comput. Syst.* 2018, *82*, 395–411.

Lin, I.-C.; Liao, T.-C. A Survey of Blockchain Security Issues and Challenges. *Int. J. Net. Secur.* 2017, *19* (5), 653–659.

Nakamoto, S. Bitcoin: A Peer-to-peer Electronic Cash System, 2008. Available: https://bitcoin.org/bitcoin.pdf

Rahulamathavan, Y.; Phan, R. C.-W.; Rajarajan, M.; Misra, S.; Kondoz, A. Privacy-preserving Blockchain Based IoT Ecosystem using Attribute-based Encryption. *11th IEEE International Conference on Advanced Networks and Telecommunications Systems*, Bhubaneswar, India, 2017, 17–20.

Siyal, A. A.; Junejo, A. Z.; Zawish, M.; Ahmed, K.; Khalil, A.; Soursou; G. Applications of Blockchain Technology in Medicine and Healthcare: Challenges and Future Perspectives. *Cryptography* 2019, *3* (1), 1–16.

Swan, M. *Blockchain-Blueprint for A New Economy*. O'reilly Publication; 2015.

Top 100 Cryptocurrencies by Market Capitalization. [Online]. Available: https://coinmarketcap.com.

Xia, Q.; Sifah, E. B.; Asamoah, K. O.; Gao, J.; Du, X.; Guizani, M. MeDShare: Trust-less Medical Data Sharing Among Cloud Service Providers via Blockchain. *IEEE Access* 2017, *5*, 14757–14767.

Zyskind, G.; Nathan, O. Decentralizing privacy: Using Blockchain to Protect Personal Data. In Security and Privacy Workshops (SPW), *IEEE,* 2015, pp. 180–184.

CHAPTER 12

Blockchain with Fault Tolerance Mechanism

SOMESH KUMAR DEWANGAN[1*], SIDDHARTH CHOUBEY[2],
ABHA CHOUBEY[2], JYOTIPRAKASH PATRA[3], and MD. SHAJID ANSARI[4]

[1]Department of CSE, GD-RCET, Bhilai, India

[2]Department of CSE, SSGI, Bhilai, India

[3]Department of CSE, SSIPMT, Raipur, India

[4]Department of CSE, RSR-RCET, Bhilai, India

[*]Corresponding author. E-mail: somdew2016@gmail.com

ABSTRACT

The blockchain network was originated from the Internet financial sector as a decentralized, immutable ledger system for transactional data ordering. Nowadays, it is envisioned as a powerful backbone/framework for decentralized data processing and data-driven self-organization in flat, open-access networks. The blockchain is a distributed, decentralized system that maintains a shared state. The blockchain is inefficient and redundant, and, that is by design, it gives us is an extreme level of fault tolerance. In particular, the plausible characteristics of decentralization, immutability, and self-organization are primarily owing to the unique decentralized consensus mechanisms introduced by the blockchain network. In this system, messages may subject to loss, damage, latency, and repetition. Also, the sending order may not necessarily be consistent with the receiving order of messages. The activities of nodes could be arbitrary, they may join and quit the network at any time; they may also dump and falsify information or simply stop working. Blockchain is a distributed ledger technology that creates a permanent, sequenced, tamper-resistant, and continuously growing list of records that is linked and secured using cryptography. With blockchain, any system of record can be replicated, shared, and synchronized across multiple locations without the need for a trusted third party for verification or authentication. While all blockchains achieve the aforementioned outcome, the architecture of block-chains depends on the use case and can range from a public permission less to a private permission design. Public blockchains, popularly represented by Bitcoin and other cryptocurrencies, are often considered to be inherently fault-tolerant by virtue of wide decentralization and methods of achieving consensus.

12.1   INTRODUCTION

A blockchain is, at heart, an integrity-focused approach to Byzantine Fault Tolerant Atomic Broadcast. The Bitcoin blockchain, for instance, uses a combination of economics and cryptographic randomization to provide a strong probabilistic guarantee that safety will not be violated, given a weak synchrony assumption, namely, that blocks are gossiped much more rapidly than they are found via the partial-hash collision lottery. A blockchain is "an incorruptible digital ledger of economic transactions that can be programmed to record not just financial transactions but virtually everything of value" (Bitcoinwiki, 2017). In practice, however, it is well known that Bitcoin's security guarantees are vulnerable to a number of subtle attacks (Robert, 2017; Bentov et al., 2016).

Blockchain has the characteristics of decentralization, stability, security, and non-modifiability. It has the potential to change the network architecture. The consensus algorithm plays a crucial role in maintaining the safety and efficiency of the blockchain. Using a right algorithm may bring a significant increase to the performance of the blockchain application, we reviewed the basic principles and characteristics of the consensus algorithms and analyzed the performance and application scenarios of different consensus mecha-nisms. We also gave a technical guidance of selecting a suitable consensus algorithm and summarized the limitations and future development of the blockchain technology. The blockchain is a distributed and decentralized system, which means that it needs to have a way of tracking the official current state of the system. It could be used for registration and issuance of digitalized assets, property right certificates, credit points, and so on. It enables transfer, payment, and transactions in a peer-to-peer way. The blockchain technology was originally proposed by Satoshi Nakamoto in a cryptography mailing list, that is, the Bitcoin. Since then, numerous applications based on the blockchain emerged, such as e-cash systems, stock equity exchanges, and smart contract systems. A blockchain system is advantageous over a traditional centralized ledger system for its full-openness, immutability and anti-multiple-spend characters, and it does not rely on any kind of trusted third party. Blockchain technology can also be used in various fields of business. One interesting implementation of blockchain technology is in the healthcare system. The method to reach consensus is highly dependent on the architecture of the network. The technique to accomplish understanding is significantly dependent on the structure of the framework. This territory will explore the differing blockchain structures, in what setting they are fitting to use and their prerequisites in that setting (Castro and Liskov, 1999).

Public blockchain

In any case, like each and every passed on structure, blockchain structures are tried with framework idleness, transmission blunder, programming bugs, security stipulations, and dull top software engineer threats. Likewise, its decentralized nature prescribes that no individual from the structure can be trusted noxious node points may develop, so does data differentiate in view of conflicting interests. To counter these potential blunders, a blockchain system needs a capable accord instrument to ensure that every center has a copy of an apparent adjustment of the total record. Standard adjustment to nonbasic disappointment instruments concerning certain issues may not be absolutely prepared for taking care of the issue that scattered and blockchain structures are looked with. A comprehensive fix to all adjustment to adaptation to noncritical failure arrangement is in need (Park et al., 2017).

12.2 VERSION OF BLOCKCHAIN

Bitcoin has come to be understood as Blockchain 1.0, a term used to describe applications that are primarily consumer-to-consumer implementations of blockchain. However, as businesses continue to examine blockchain's potential, Blockchain 2.0 and 3.0 use cases are beginning to emerge (Bitcoinwiki, 2017; Lamport, 2001).

Blockchain 1.0

Use cases are public, anonymous, and fully transparent, with Bitcoin being the flagship example. Blockchain 1.0 often connotes consumer-to-consumer transactions (i.e., currency) (Nakamoto, 2008) in a decentralized network relying on large numbers of specialized participants for transaction verification.

Blockchain 2.0

Applications are permissioned enterprise-to-enterprise implementations of blockchain, understood to involve elements of smart contracting. Ethereum has often been cited as Blockchain 2.0 when compared with Bitcoin, given its improved flexibility for creating decentralized applications and smart contracting capabilities.

Blockchain 3.0

Describe implementations that leverage the convergence of AI. Blockchain and IoT to allow users to monetize the data collected by internet-connected devices—ranging from cars to mobile base stations—for sale to organization interested in buying it. Rather than being consumer-to-consumer or enterprise-to-enterprise, Blockchain 3.0 use cases encompass transactions and exchange between devices and between consumers and enterprises (Pommier, 2017).

12.2.1   VALUE TO BUSINESS

Blockchain network feature multiple core characteristics that drive value including decentralization, immutability, consensus, integrity, transparency, and fault tolerance. When blockchains are made private and suitable for business applications, these characteristics can together support secure, reliable, and efficient blockchain networks for enterprises of all sizes (Bitcoinwiki, 2017).

12.2.2   DECENTRALIZATION

Ownership of information is decentralized and shared across a peer-to-peer network and continually reconciled. Blockchain can eliminate the need to rely on trusted third parties or intermediaries to verify transactions, maintain record, and attest to the authenticity of information (Schwarz, 2014).

12.2.3   IMMUTABILITY

Data kept on the blockchain is append-only and cannot be deleted or changed once a transaction has occurred.

12.2.4 TRANSPARENCY

All participants in the blockchain have visibility into a complete record of transactions. In the case of private or enterprise blockchains, there can be modifications to this characteristic based on the use case and the type of blockchain platform used (Bitcoinwiki, 2015).

12.2.5 INTEGRITY

Transactions are executed exactly as specified and data is accurate, preserving the integrity of the information.

12.2.6 CONSENSUS

Participating parties must give consensus before a new transaction is added to the network with transactions executed exactly as specified.

12.2.7 FAULT TOLERANCE

In public blockchain networks, since the ledger is replicated, shared, and synchronized across participants, failure of one or more component of the network does not cause failure of the entire system (Bitcoinwiki, 2016). All parties have visibility into the complete record of transactions. Highly distributed public networks are fault-tolerant by design, though private enterprise blockchain networks consisting of fewer nodes are made fault-tolerant by the quality of their underlying and surrounding infrastructure, and special services involved in consensus, transaction validation, or other similar tasks. Blockchain's characteristics make it a more attractive option than many next best alternatives. Although businesses today rely heavily on brokers, lawyers, notaries, and conventional databases, these systems and intermediaries are often time-consuming, expensive, cumbersome, and comparatively inefficient methods of maintaining trust in systems of record and the exchange of value between parties (Caldwell, 2017).

12.3   BLOCKCHAIN SYSTEM TAXONOMY

Blockchain implementations can be architected in several different ways. Roughly three different categories exist: public, private, and consortium blockchains (Bitcoinwiki, 2017). This section will explore the different blockchain architectures, in what setting they are appropriate to use and their constraints in that setting.

12.3.1 PUBLIC BLOCKCHAIN

In a public blockchain, depending on the design, users can get rewards based on their participation like in the Bitcoin blockchain (Bitcoin Forum, 2011; Bitcoinwiki, 2017). This architecture is usually referred to as a permission less blockchain. In a permission less blockchain, there is no single owner since all users maintain a copy of the ledger locally. It is considered to be "fully decentralized" (Bitcoinwiki, 2017; Pommier, 2017).

The underlying rule is that a user can only participate in the consensus process equivalent to the economic resources that they possess, either computational (PoW) or stake (PoS). The limitations of a public architecture refers to many aspects, one important one being the symmetry of information. Since all the information on the ledger is symmetric, the use of a public blockchain in a business scenario can become troublesome (Bitcoinwiki, 2017).

12.3.2   CONSORTIUM BLOCKCHAIN

Consortium blockchains are normally limited to the companies involved in the business scenario or a part of it. These blockchains are consequently permissioned and are usually considered partially decentralized not only because of the smaller number of nodes compared to public blockchains, but also because of the hierarchy in the nodes. The hierarchy refers to the fact that different firms usually have different views of the data that helps securing loss of business intelligence information to competitors (Bitcoinwiki, 2017). Consortium blockchains usually don't require the extensive security that mainstream consensus algorithms such as proof-of-work provide since the members are trusted (Bitcoin forum, 2011). This is a disadvantage compared to public blockchains where anyone can join because deciding who can join the network or who can commit data to the ledger is usually done in a centralized manner (Bitcoinwiki, 2017; Larimer, 2014).

12.3.3   PRIVATE BLOCKCHAIN

This blockchain type is very similar to consortium blockchains but instead with a single organization as a participant. Many argue that private blockchains add no value and that a normal database can be used instead since there is no need to trust other parties and exchange of information between them. Gideon Greenspan has defined a set of criteria that need to be fulfilled for blockchain to be a good use case (Bitcoinwiki, 2014). The second criteria state that there has to be multiple writers in the database, which means more than one entity is required (Bitcoinwiki, 2014).

12.4   FAULT TOLERANCE

Adaptation to noncritical failure is a framework that is dependent to the disappointment of components inside the framework. It likewise might be known as a safeguard structure. Fault resilience is a framework that is dependent to the disappointment of components inside the framework. It likewise might be known as a safeguard plan (Multichain, 2015).

A fault-tolerant framework may continue working okay, after one of the power supplies misses the mark; other systems may have a limp home condition, empowering the structure to save fundamental data or empowering you to drive to an ensured spot to supplant a punctured tire. Correspondence, banking, air terminal guideline, transportation, and various diverse fields have systems where a failure to work may provoke disastrous result. Making a system that may experience fragment, subsystem, or programming disillusionments and the structure can continue with movement (Greenspan, 2015).

Fault tolerance: Characteristics

A fault-tolerant framework may have at least one of the accompanying characteristic. A fault-tolerant structure may have no less than one of the going with trademark.

- **No single point of failure**
  It suggests if a capacitor, square of programming code, a motor, or any single thing flops then the system does not fall flat. As to we have a precedent, may medical clinics have reinforcement control frameworks in the event that the network control flops, in this manner keeping basic framework inside the emergency clinic operational. Basic frameworks may have numerous excess plans to keep up an

abnormal state of adaptation to noncritical failure and flexibility (Sompolinsky and Zohar, 2013).

- **No single point repair takes the system down**
  Extending the single point failure idea thought figured, influencing a fix of a failed part does not require closing down the structure. Consequently, it in like manner infers the system remain on the web and operational in the midst of fix. This may show challenges for both the design and the help of a structure. Hot swappable power supplies are an instance of a fix movement that keeps the structure working while in the meantime displacing a faulty power supply (Eyal and Sirer, 2014).

- **Fault isolation or identification**
  The framework can recognize when a blame happens inside the framework and does not allow the defective component to unfavorably impact to useful ability. The broken component are recognized and detached.

- **Fault containment**
  At the point when disillusionment happens it may result in harm to different components within the framework, in this way making a second or third fault and framework dissatisfaction (Sameeh, 2016). For instance, if a basic circuit crashes and burns it may extend the current over the system hurting reason circuits powerless to withstand high current conditions. The idea of accuse control is to avoid or restrict unintentional aftereffect caused by a solitary point disillusionment

- **Robustness or variability control**
  At the point when framework experiences a lone point disappointment, the framework changes. The change may cause transient or perpetual changes influencing how the working components of the structure response and capacity. Assortment occurs, and when a failure occurs there consistently is an expansion in capriciousness (Schwarz, 2014).

- **Reversion state operation (fall back or limp-along)**
  There are various ways a structure may modify its execution when a disappointment occurs, empowering the system to continue working in some style. For example, if part of a PC's cooling structure misses the mark, the central processor unit may lessen its speed or request execution rate, feasibly lessening the warmth the CPU creates. The miss the mark disillusionment causes lost cooling limit and the CPU

modify to suit and goes without overheating and coming up short. Here and there, the system may be prepared to manage with no or only inconsequential loss of utilitarian capacity or the inversion assignment fundamentally restricts the structure movement to a basic few.

• **Enterprise blockchain and the need for fault-tolerance**
  Enterprises are deeply conscious about the privacy and security of their data. They also want to shield their intellectual property employed in business processes. Most enterprises do not want their transactions and business processes to be visible to anyone other than vetted and authorized identities. It may be ideal for every transaction to be visible in a public blockchain; however, such transparency is often not conducive to business, ranging from competitive to regulatory reasons (Bitcoinwiki, 2015).

Enterprise blockchains warrant consortium-based business models with known, vetted participants and a limit on the number of nodes involved in consensus for reasons of performance and deterministic results. As a result, there is a strong correlation between the number of nodes involved in consensus and the level of fault-tolerance required for each node in a private, enterprise blockchain. A public, permission less blockchain can tolerate the failure of any given node or a group of nodes because of the Byzantine fault tolerance principles of a very large network and consensus established by the remaining nodes. However, given the more consolidated nature of enterprise blockchains across a relatively small number of nodes, node failure can compromise the blockchain, since each enterprise is highly dependent on its node(s) for participating in the ecosystem. Consensus and hence entire transactions between enterprises are at stake due to a node failure. In order to assure continuous availability and business resiliency, enterprise blockchains commonly exhibit a higher need for fault-tolerance on a per node basis (Pommier, 2017).

12.4.1 FAULT TOLERANCE IN THE BLOCKCHAIN

The blockchain is a distributed, decentralized system that maintains a shared state. While consensus algorithms are designed to make it possible for the network to agree on the state, there is the possibility that agreement does not occur. Fault-tolerance is an important aspect of blockchain technology. The blockchain is inefficient and redundant, and that is by design. That's what gives us immutability. And another thing it gives us is an extreme level of fault-tolerance. At its heart, blockchain runs on a peer-to-peer network architecture in which every node is considered equal to every other node. And unlike traditional client–server models, every node acts as both a client and a server (Bitcoinwiki, 2014). Figure 12.1 shows fault tolerance mechanism in blockchain.

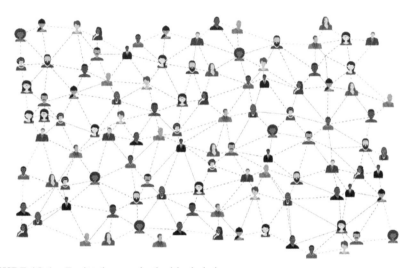

FIGURE 12.1   Fault tolerance in the blockchain.

And so, we continue this redundancy down at the network level, where we're asking all these nodes to perform the same work as all these other nodes. Like any peer-to-peer system, we have an extremely high degree of fault-tolerance. In fact, if we have two or more nodes online in a blockchain system, we still have a working blockchain. And when you think about that amazing fact given the scale of major public blockchains, you can see the inbuilt fault-tolerance (Sameeh, 2016).

Let's think about Bitcoin for an example

That's a blockchain that consists of over thirty thousand nodes coming to a consensus on every block. As long as we have two or more of those nodes online and are able to communicate, we still have a working solution that gives us a tremendous margin for error, for nodes coming and going offline, for network transport issues, and it makes blockchain really a great platform to use in environments with less than ideal networks and power infrastructure. Because we can have nodes come offline, go back online and when a node comes back online after being offline for a while, all it has to do is sync up, and get all the data that it missed while it's been offline from all of its peers, and then it's right back online participating like all the rest. This is very different from the centralized systems that blockchain aims to replace. In a traditional client–server model, if that server is offline, those clients have no way of getting the data that they requested or performing the operations they'd like to perform (CoinDesk Inc., 2014).

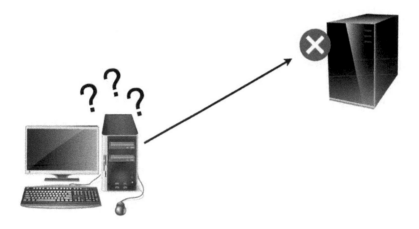

This is not the case in blockchains

And if we look back historically at other peer-to-peer solutions, solutions such as BitTorrent or Napster, we've seen the tremendous difficulty that authorities have had taking some of these networks offline. That is because of the adaptation to noncritical failure you get from a shared architecture. In reality, we saw this recently amid the Arab Spring, when the Egyptian government chose one night to close down Internet access for the whole nation. Well, within 24 hours Egypt was back online and connected to the Internet through a network-sharing mechanism known as mesh networking, which at its heart is just a peer-to-peer method for sharing Internet connectivity. So, we know that peer-to-peer has a long history of providing extremely high fault tolerance and reliability, and that's why it's been chosen to build a platform like a blockchain on top of it. So, if you're looking for a solution platform that offers you that kind of incredible fault tolerance, if you're looking to deploy a solution into areas with less than ideal infrastructure or under conditions where nodes may come online and go offline frequently, then blockchain may be a really good platform to look at (Robert, 2017).

12.5   CONSENSUS

Consensus is the word for a general agreement or majority of opinion between a number of subjects (Cachin, 2016). It is a fundamental part of everyday life like deciding where to eat with a group of friends or which movie to watch. In a distributed system, deciding if data is accurate when the system is updated is a complex task. William Mougayar describes this decentralized scheme interestingly: "At the heart of understanding the severity of the blockchain paradigm shift lies the basic understanding of the concept of decentralized consensus" (Caldwell, 2017). Consensus in a blockchain network refers to the process where the distributed nodes agree on the history and the final state of the data on the ledger, usually referred to as distributed consensus (Bitcoin forum, 2011). Since all participants in the network hold the data, they can also be a part of the decision-making (Vasin, 2014).

Byzantine fault tolerance (BFT)

The Byzantine Generals' Problem was distributed in 1982 by Lamport, Shostak, and Pease, to guarantee that failing parts that give clashing data are dealt with inside a PC framework. BFT is the characteristic of a system that handles the different types of faults that belongs to the Byzantine Generals' Problem. Having BFT is not solely used on the blockchain itself. However, in the consensus algorithms explored below, they are explained in relation to the blockchain. Byzantine faults are the most serious and difficult to manage. BFT has been required in plane motor frameworks, atomic power plants, and practically any framework whose activities rely upon the consequences of a lot of sensors. Indeed, even SpaceX was thinking about it as a potential prerequisite for their systems. The trademark is known as "Byzantine adaptation to non-critical failure." BFT is one of those ideas worth comprehension (Ripple. Available: from https://ripple.com/). The capacity to endure what PC researchers call "byzantine disappointments" is an essential piece of blockchains' capacity to keep up dependable records of exchanges in a straightforward, carefully designed way. The representation depicts an issue that plagues numerous PC systems. At the point when a gathering is endeavoring to settle on an aggregate choice about how it will act, there is a hazard that swindlers inside the gathering may send blended messages about their inclinations. The swindlers may disclose to certain individuals from the gathering that they wish to complete a certain something, and tell different individuals from the gathering the inverse. In any disseminated figuring condition, which means a domain where different clients, applications, servers, or different sorts of hubs create nature (like a blockchain), there is a hazard that rebel or temperamental on-screen characters could make the earth break separated. A server group won't function admirably if a few servers inside it neglect to pass information reliably to different servers. A PC system will come up short if the gadgets on it don't concede to a typical systems administration convention to utilize while trading data. The calculation referenced in the past area is BFT as long as the quantity of back-stabbers doesn't surpass 33% of the commanders. Different varieties exist that make taking care of the issue less demanding, including the utilization of computerized marks or by forcing correspondence limitations between the companions in the system, maybe no place is BFT more fundamental than on a blockchain. Most customary disseminated registering conditions have focal arrangement databases or experts that can help right wrongs if Byzantine disappointments happen. In any case, on a blockchain there is, by definition, no focal expert. Blockchains' capacity to genuine exchanges dependent on network agreement alone is the thing that makes them so amazing. This substantial dependence on network accord additionally makes Byzantine blame an especially vital test for blockchain. On the off chance that a few individuals from the network send conflicting data to others about exchanges, the dependability of the blockchain separates, and there is no expert that can venture in to address it. In this way, except if you can put outright trust in everybody who takes part in your blockchain (Nxt wiki, 2016).

Consensus protocol

A working consensus protocol preserves the quality of the data on the ledger since strict rules apply when updating the ledger. Consensus protocols can be formally analyzed by focusing on certain design parameters such as node identity management, consensus finality, scalability (number of nodes participating in consensus), performance (latency, transactions per second), adversarial behavior tolerance, network assumptions, and lastly existence of correctness proofs (Bitcoinwiki, 2017; Ethereum).

Proof of work (PoW)

Block validators have to prove that they have performed some amount of computational work, which is done by finding a hash value that is less than the difficulty level set by the network (CoinDesk Inc., 2014). Bitcoin's difficulty is set according to the total hash power of the network to create a new block every ten minutes. Solving the PoW puzzle and finding the value is also known as mining (CoinDesk Inc., 2014). The first node that finds the value lower than the difficulty wins the mining race and gets its proposed block added to the blockchain along with a mining reward that is paid in Bitcoin (CoinDesk Inc., 2014). Because the network is fully distributed and asynchronous, temporary forks can occur where more than one validator finds a winning hash value. This means that each validating node adds the proposed block to the blockchain and broadcasts it to the network, thus creating diverging branches. The PoW consensus model is well suited for environments where no authentication of participants is needed and where anyone can join as a validator. PoW is, however, vulnerable to dominance attacks, or more commonly known as 51% attacks.

Another problem with PoW is the slow confirmation rate, which is around seven transactions per second (Bitcoinwiki, 2017). Compared to VISA's 10000 transactions per second, this throughput is far from usable for commercial purposes (CoinDesk Inc., 2014). Since mining requires computational work, mining gear is needed to participate in the validation of blocks. As Nakamoto stated in the paper, nodes vote using their CPU. However, actually getting a reward using CPU mining today is not feasible with the vast amount of competition. PoW exploits the scarcity of computational resources by choosing a problem that can only be solved by guessing. There is no limit on the number of guesses that a miner can make at once. So, PoW incentives miners to run as many mining machines as possible to maximize the probability that they are the first to find a solution to the problem. Since mining computers take money to purchase and money to run, the amount of control that a user can exert over the blockchain network is limited by the amount of money they have available to invest in mining equipment.

The security of the PoW consensus is based on the assumption that no one controls more than half of the computational resources of a blockchain's mining network.

The idea of putting coins to be "staked" prevents bad actors from making fraudulent validations—upon false validation of transactions, the amount staked will be forfeited. Hence, this incentivizes forgers to validate legitimately. In the recent year, PoS has gained attention, with Ethereum switching toward a PoS from a PoW consensus system.

PoW can support a working consensus with at most 25% of network power acting maliciously (Bitcoinwiki, 2017). Up until recently, the understanding was that PoW could handle everything below 50% and work properly but that was lowered to 25% after further research (Schwartz et al., 2014).

Proof of stake (PoS)

The disadvantages of PoW have sparked interest in designing algorithms that don't have nearly as high electricity consumption in the mining process. PoS replaces the computational work in the mining process with using the validator's stake/ownership of a certain cryptocurrency. Instead of spending money on electricity and mining gear, validators can instead buy crypto-currency and use that currency to engage in the validating process to get rewards and keep the network operating. The amount of currency at stake by the validator corresponds to their probability of being chosen as a validator (CoinDesk Inc., 2014).

The security is not achieved by consuming electricity, but instead by putting up economic value in the form of a stake in cryptocurrency, which can be lost by malicious acts by the validator. Subsequently, PoS is more penalty-focused contrasted with PoW that is compensate-based to guarantee security, (Robert, 2017) removes the vitality and computational power necessity of PoW and replaces it with stake. Stake is alluded to as a measure of cash that a performing artist is happy to bolt up for a specific measure of time. Consequently, they get an opportunity relative to their stake to be the following head and select the following square. There are different existing coins that utilize unadulterated PoS, for example, Nxt and Blackcoin. Figure 12.2 shows proof of stake.

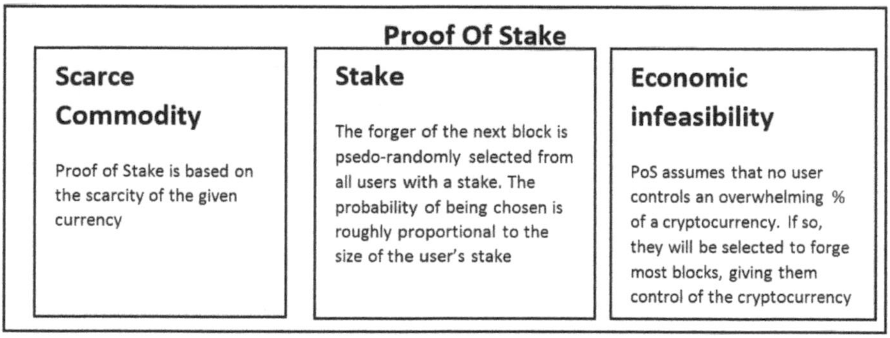

Proof Of Stake		
Scarce Commodity Proof of Stake is based on the scarcity of the given currency	**Stake** The forger of the next block is psedo-randomly selected from all users with a stake. The probability of being chosen is roughly proportional to the size of the user's stake	**Economic infeasibility** PoS assumes that no user controls an overwhelming % of a cryptocurrency. If so, they will be selected to forge most blocks, giving them control of the cryptocurrency

FIGURE 12.2   Proof of stake.

Delegated proof of stake (DPoS)

DPoS is similar to PoS in regard to staking but has a different and a more democratic system that is said to be fair. Like PoS, token holders stake their tokens in this consensus protocol. Instead of the probabilistic algorithm in PoS, token holders within a DPoS network are able to cast votes proportional to their stake to appoint delegates to serve on a panel of witnesses—these witnesses secure the blockchain network. In DPoS, delegates do not need to have a large stake, but they must compete to gain the most votes from users (Echevarria, 2017). It provides better scalability compared to PoW and PoS as there are fully dedicated nodes who are voted to power the blockchain. Block producers can be voted in or out at any time, and hence the threat of tarnishing their reputation and loss of income plays a major role against bad actors. No doubt, DPoS seem to result in a semi-centralized network, but it is traded off for scalability. Like PoS, DPoS has also gained attention over the years with several projects adopting this consensus algorithm. Since it was invented by Dan Larimer, DPoS has been refined continuously, from BitShares to Steem and now in EOS (Bitcoinwiki, 2017).

Proof of authority (PoA)

PoA is known to bear many similarities to PoS and DPoS, where only a group of preselected authorities (called validators) secure the blockchain and are able to produce new blocks. New blocks on the blockchain are created only when a super majority is reached by the validators (Nakamoto, 2008).

The identities of all validators are public and verifiable by any third party—resulting in the validator's public identity performing the role of proof of stake. As these validators identity are at stake, the threat of their identity being ruined incentivizes them to act in the best interest of the network. Due to the fact that PoA's trust system is predetermined, concerns have been raised that there might be a centralized element with this consensus algorithm. However, it can be argued that semi-centralization could actually be appropriate within private/consortium blockchains—in exchange for better scalability. Newer blockchain startups has ventured into implementing PoA. In addition, Ethereum test nets such as Rinkeby and Kovan explore the use of a PoA consensus algorithm.

The Byzantine Generals' problem

BFT is so named on the grounds that it speaks to an answer for the "Byzantines general concern," an intelligent difficulty that scientists Leslie Lamport, Robert Shostak, and Marshall Pease portrayed in a scholarly paper published in 1982. Basically, it envisions a gathering of Byzantine officers and their armed forces encompassing a manor and getting ready to assault. To be effective, these armed forces should all assault in the meantime. Be that as it may, they realize that there is a swindler in their middle. The issue they face is one of propelling a fruitful assault with one, obscure awful on-screen character in their framework.

Potential solution

Blockchains are decentralized records which, by definition, are not obliged by a central expert. As a result of the regard set away in these records, horrendous on-screen characters have immense money-related inspirations to endeavor and cause inadequacies. Everything considered, BFT, and along these lines a response for the Byzantine Generals' Problem for blockchains is much needed. In the nonappearance of BFT, a companion can transmit and post false trades enough discrediting the blockchain's constancy. To intensify the circumstance, there is no central master to rule and fix the damage. The tremendous accomplishment when Bitcoin was planned, was the use of PoW as a probabilistic response for the Byzantine Generals' Problem as depicted through and through by Satoshi Nakamoto in this email.

This substantial dependence on network accord likewise makes Byzantine blames an especially imperative test for blockchain. On the off chance that a few individuals from the network send conflicting data to others about exchanges, the unwavering quality of the blockchain separates, and there is no specialist that can venture in to address it. In this way, except if you can put outright trust in everybody who takes an interest in your blockchain (which you can't most of the time), you need an approach to secure against the Byzantine blames that could happen if a few individuals appropriate wrong, deceptive or pernicious exchange data.

BFT is a significant bit of a successful blockchain and there are diverse habits by which it might be realized. Using that approach to manage take requires gauging the nature and necessities of the system related with the blockchain. The responses for BFT that have made structures like Bitcoin conceivable may not work commendably in the blockchain employments of things to come.

Algorithm

Our calculation ensures security similarly as convenience. With mixed up center points in the consensus making near $[(n-1)/3]$, the value and dauntlessness of the system is guaranteed. In it, $n = |R|$ proposes without a doubt the quantity of center points joined in the consensus making while R speaks to the game plan of accord centers. Given $f = (n-1)/3$, f speaks to the most extraordinary number of inaccurate enter points allowed in the framework. In reality, the outright record is kept up by bookkeeping center points while conventional centers don't participate in the understanding making. This is to demonstrate the entire accord making procedures (Park et al., 2017). All understanding center points are required to keep up a state table to record current accord status. The instructive gathering used for an understanding from its begin to its end is known as a View. We recognize each View with a number v, starting from 0 and it may augment till achieving the accord. We recognize each understanding center point with a number, starting from 0, the prop up hub is numbered n − 1. For each round of understanding making, a center point will play speaker of the house while distinctive centers play congressmen. The speaker's number p will be determined by the going with computation: Hypothetically the present square stature is h, then $p = (h - v)$ *modn, p's* regard range will be $0 \leq p <$ (King, 2013).

Another square will be made with each round of accord, with at any rate $n - f$ signatures from bookkeeping center points. Upon the age of a square, another round of consensus making will begin, resetting $v = 0$.

12.6   GENERAL PROCEDURE

Set the time interims of square age as t, under ordinary conditions, the calculation executes in the accompanying procedures

1. A node communicates exchange information to the whole system, joined with the senders' signature.
2. All accounting hubs screens exchange information broadcasting freely and stores the information in its memory individually.
3. After the time t, the speaker sends ⟨*PrepareRequest,h,v,p,block,*⟨*block*⟩*σp*⟩.
4. After accepting the proposition, congressmen I send ⟨*PrepareResponse,h,v,i*⟨*block*⟩*σi*⟩
5. Any node, after accepting at any rate $n - f$⟨*block*⟩*σi*⟩, achieves an agreement and distributes a full square.
6. Any node, in the wake of accepting the full square, erases the exchange being referred to from its memory and starts the following round the accord (Caldwell, 2017).

It is required that, for all the agreement hubs, in any event $n - f$ hubs are in a similar unique state. This is to state, for every one of the hubs I, the square tallness h and View number v are the equivalent. This isn't troublesome, consistency of h could be come to by synchronizing the squares while consistency of v could come to by changing the View. Square synchronizing isn't canvassed in this article. For View change, check next segment.

Nodes, subsequent to checking the telecom and accepting the proposition, will approve the exchanges. They can't compose an unlawful exchange in the memory once the last is uncovered. In the event that an illicit exchange is contained in the proposition, this round of accord will be surrendered and the View change will occur right away. The approval methods are as per the following:

1. Is the information arrangement of the exchange steady with the framework rules? Assuming no, the exchange is ruled unlawful.
2. Is the exchange as of now in the blockchain? On the off chance that indeed, the exchange is ruled illicit.
3. Are all the agreement contents of the exchange accurately executed? Assuming no, the exchange is ruled illicit.
4. Is there different spend in the exchange? On the off chance that indeed, the exchange is ruled unlawful.
5. If the exchange had not been ruled illicit in the above-mentioned methodology, it will be ruled lawful.

Figure 12.3 Shows flowchart.

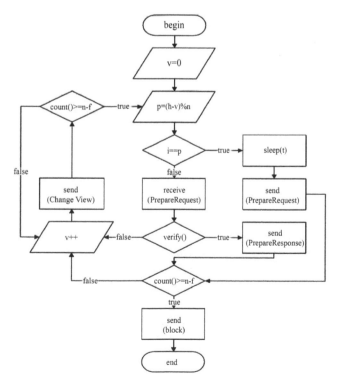

FIGURE 12.3 Flowchart (algorithm).

In the event that, after $2v+1 \cdot t$ time interim, the hubs I can't achieve an agreement or should they get recommendations that contain unlawful exchanges, the View Change will happen:

1. Given k = 1, $vk = v +$
2. Nodes I send View Change ask for <*ChangeView, h, v, i, vk*>
3. Once any hub gets at any rate $n - f$ same vk from various I, the View Change is finished. Set $v = vk$ and the agreement making starts
4. If, after $2v+1 \cdot t$ time interim, the View Change isn't finished, the k will increment and back to stage 2.

With the k expanding, the additional time holding up time will increment exponentially, so visit View Change will be dodged and hubs are encouraged to achieve consistency over (Bitcoinwiki, 2017).

Prior to the finishing of View Change, the first View v is as yet substantial, so pointless View Change brought about by intermittent system inactivity can be stayed away from.

Fault tolerance capacity

Our calculation gives $f = [(n-1)/3]$ adjustment to inside inability to an understanding system that incorporates n center points. This obstruction limit consolidates security and convenience and is suitable for any framework condition.

Demand information from hubs contain sender marks, so noxious accounting hubs can't distort demands. Rather, they will attempt to turn around the framework status back to the past, driving the framework to fork.

Hypothetically, in the framework state of the system, understanding center points are isolated into three areas: $R = R1 \cup R2 \cup$, and $R1 \cap R2 = \emptyset$, $R1 \cap F = \emptyset$, $R2 \cap F = \emptyset$. Also hypothetically, both $R1$ and $R2$ are clear bookkeeping center points in an information storage facility that they can simply talk with centers in their set; F are in general poisonous center points in coordination; what's more, the framework territory of F empowers them to talk with any center point, including $R1$ and $R2$. In case F wishes to fork the system, they have to accomplish accord with $R1$ and disperse squares, and thereafter accomplish a second understanding without teaching the $R2$, denying the concurrence with $R1$. To accomplish this, it is essential that $|R1| + |F| \geq n - f$ and $|R2| + |F| \geq n -$. In the most desperate result believable, $|F| = f$, for instance the amount of noxious center points is at the most extraordinary that the system could bear the recently referenced association advances toward getting to be $|R1| \geq n - 2f$ and $|R2| \geq n - 2f$. Included, $|R1| + |R2| \geq$

$2n - 4f$, which could be revamped as $n \leq 3f$. Given that $f = [\ (n-1)/3\ ]$, which denies with the past, it will in general be exhibited that the structure can't be forked inside the opposition broaden. (Bitcoinwiki, 2017).

Practical Byzantine fault tolerance

The Practical Byzantine fault tolerance (PBFT) calculation was created by Castro and Liskov (Elliptic curve digital signature algorithm, 2017) to tackle the issue of achieving accord even with Byzantine hubs. It was the principal common sense execution of Byzantine blame tolerant agreement calculations.

PBFT will use state–machine replication and copy in favor of changing the main system. It should be validated in PBFT to arrange all events and approval through numerous rounds. PBFT is utilized in private settings where no verification is required. Attributes that enhance the correspondence are marking and scrambling messages among copies and customers, and limiting the quantity of traded messages. This is to make the calculation usable despite Byzantine flaws. It requires 3f + 1 reproductions to endure f Byzantine hubs (Elliptic curve digital signature algorithm, 2017). This can likewise be detailed as under 33% of the hubs can be defective for accord to be come to. Malignant assaults and programming mistakes are progressively normal. The developing dependence of industry and government on online data administrations makes noxious assaults increasingly alluring and makes the outcomes of effective assaults progressively genuine. What's more, the quantity of programming mistakes is expanding because of the development in size and unpredictability of programming. Since vindictive assaults and programming blunders can make flawed hubs show Byzantine (i.e., discretionary) conduct, Byzantine-blame tolerant calculations are progressively vital.

Advances on consensus algorithm

A decent accord figuring suggests profitability, safety, and settlement. As of late, different endeavors have been made to improve ascension estimations in blockchain. New accord counts are prepared proposing to deal with some particular issues of blockchain. The guideline thought of PeerCensus (Bentov et al., 2016) is to decouple square creation and trade attestation with the objective that the understanding velocity can be in a general sense extended (Vasin, 2014). Furthermore, Kraft (Bentov et al., 2014) proposed another agreement strategy to ensure that a square is delivered in a for the most part enduring pace. It is realized that high squares age rate deal Bitcoin's security.

So the Greedy Heaviest-Observed Sub-Tree (GHOST) chain assurance rule (Eyal and Sirer, 2014) is proposed to take care of this issue. As opposed to the longest branch contrive, GHOST stacks the branches and excavators could pick the better one to seek after. Chepurnoy et al. (Eyal et al., 2014) presented another ascension figuring for companion to peer blockchain structures where any individual who gives noninteractive confirmations of retrievability to the past state previews is agreed to create the square.

Future possibilities

In addition to those use cases that have been demonstrated in proofs of concept and pilot studies to date, many other Blockchain 2.0 and 3.0 possibilities are gaining industry backing and expected to be areas of focus in the near future. The Toyota Research Institute has announced activities in trialing blockchain for the sharing of driving and autonomous vehicle-testing data. In the telecommunications industry, the Mobile Authentication Taskforce, which consists of AT&T, Verizon, and other network operators, aims to protect enterprises and consumers from identity theft, fraud, and data theft. Beyond identity management and authentication applications, telecom equipment manufacturers are exploring the use of blockchain as a means to enable future data monetization in smart cities. Nokia has proposed plans to create a sensors- as- a- service network powered by blockchain for the monetization of live environmental data—including weather and traffic—collected for smart cities (Duong et al., 2016).

12.7   CONCLUSION

Blockchain can provide substantial benefits to businesses from cost savings to the creation of new business models and platforms of exchange. The blockchain is distributed, decentralized system that maintains a shared state. The blockchain is inefficient and redundant, and that is by design it gives us is an extreme level of fault tolerance. However, private blockchain networks for Blockchain 2.0 and 3.0 use cases are only as strong as their weakest links. With blockchain, any system of record can be replicated, shared, and synchronized across multiple locations without the need for a trusted third party for verification or authentication. Servers that provide robust fault tolerance for demanding enterprise applications can help to ensure the overall value of blockchain across industries. Public blockchains, popularly represented by Bitcoin and other cryptocurrencies, are often considered to be inherently fault tolerant.

KEYWORDS

- **cryptocurrency**
- **hash**
- **encryption**
- **node**
- **forks**

REFERENCES

Activity. 2017 (Online). Available: https://eprint.iacr.org/2017/367.pdf. Atlanta, GA, 2014, pp. 355–362. delegated-proof-of-stake-consensus/. Networked Systems Design and Implementation, Berkeley, CA, 2016; pp. 45–59.

Armknecht, F.; Karame, G. O.; Mandal, A.; Youssef, F.; Zenner, E. Ripple: Overview and Outlook. In . Springer Cham: Switzerland, 2015; pp. 163–180. Ripple (Online). Available: from https://ripple.com/.

Back, A. Hashcash: A Denial of Service Counter-Measure. 2002 (Online). Available: ftp://sunsite.icm. edu.pl/site/replay.old/ programs/ hashcash/ hashcash .pdf.

Bastiaan, M. Preventing the 51%-Attack: A Stochastic Analysis of Two Phase Proof of Work in Bitcoin. 2015 (Online). Available: http://referaat.cs.utwente.nl/conference/22/paper/7473/preventingthe-51-attack%20-astochastic- analysis-of-two-phase-proof-of-work-in-bitcoin.pdf.

Ben-Or, M.; Linial, N. Collective Coin Flipping. Institute of Mathematics and Computer Science. The Hebrew University, Jerusalem, Israel, 1990.

Bentov, I.; Gabizon, A.; Mizrahi, A. Cryptocurrencies without Proof of Work. In . Springer: Heidelberg, 2016; pp. 142–157.

Bentov, I.; Lee, C.; Mizrahi, A.; Rosenfeld, M. Proof of Activity: Extending Bitcoin's Proof of Work via Proof of Stake. (3), 34–37, 2014.

Bessani, A.; Sousa, J.; Alchieri, E. E. P. In Proceedings of 2014 44th Annual IEEE/IFIP International Conference on Dependable Systems and Networks.

Bitcoin forum. Topic: Proof of Stake Instead of Proof of Work. 2011 (Online). Available: https://bitcointalk.org/index.php?topic=27787.0.

Bitcoin Stack Exchange. Can Someone Explain How the Bitcoin Blockchain Works? 2017 (Online). Available: https://bitcoin.stackexchange.com/questions/12427/can-someone-explain- how-the-bitcoinblockchain-.

Bitcoinwiki. Block Hashing Algorithm. 2015 (Online). Available: https://en.bitcoin.it/wiki/Block_hashing_algorithm.

Bitcoinwiki. Genesis Block. 2017 (Online). Available: https://en.bitcoin.it/wii/Genesis_block.

Bitcoinwiki. Irreversible Transactions. 2017 (Online). Available: https://en.bitcoin.it/wiki/Irreversible_Transactions.

Bitcoinwiki. Proof of Stake. 2014 (Online). Available: https://en.bitcoin.it wiki/Proof_of_Stake.

Bitcoinwiki. SHA-256. 2016 (Online). Available: https://en.bitcoin.it/wiki/SHA-256.

Bitcoinwiki. Weaknesses. 2017 (Online). Available: https://en.bitcoin.it/wiki/Weaknesses# Attacker_has_a_lot_of_computing_power.

Blocki, J.; Zhou, H. S. Designing Proof of Human-Work Puzzles for Cryptocurrency and Beyond. In . Springer: Heidelberg, 2016; pp. 517–546.

Cachin, C. In . Proceedings of ACM Workshop on Distributed Cryptocurrencies and Consensus Ledgers, Chicago, IL, 2016.

Caldwell, C. K. Cunningham Chain. 2017 (Online). Available: http://primes.utm.edu/glossary/xpage/CunninghamChain.html.

Castro, M.; Liskov, B. Byzantine Fault Tolerance. U.S. Patent 6671821 B1, Dec 30, 2003.

Castro; M.; Liskov, B. In. Proceedings of the Third Symposium on Operating Systems Design and Implementation, New Orleans, LA, 1999,; pp. 173–186.

Chepurnoy, A.; Duong, T.; Fan, L.; Zhou, H. S. TwinsCoin: A Cryptocurrency via Proof-of-Work and Proof-of-Stake. 2017 (Online). Available: https://eprint.iacr.org/2017/232.pdf.

CoinDesk Inc., What are Bitcoin Mining Pools? 2014 (Online). Available: https://www.coindesk.com/information/get-started-mining-pools/.

Duan, S.; Meling, H.; Peisert, S.; Zhang, H. BChain: Byzantine Replication with High Throughput and Embedded Reconfiguration. In. Springer: Cham, Switzerland, 2014; pp. 91–106.

Duong, T.; Fan, L.; Zhou, H. S. 2-Hop Blockchain: Combining Proof-of-Work and Proof-of-Stake Securely. 2016 (Online). Available: https://eprint.iacr.org/2016/716.pdf.

Dziembowski, S.; Faust, S.; Kolmogorov, V.; Pietrzak, K. Proofs of Space. In. Springer: Heidelberg, 2015; pp. 585–605.

Echevarria, R. The Second Coming of Blockchain. 2017 (Online). Available: https://software.intel.com/enus/blogs/2017/02/14/the-second-coming-of-blockchain.

Elliptic Curve Digital Signature Algorithm. 2017 (Online). Available: https://en.bitcoin.it/wiki/Elliptic_Curve_Digital_Signature_Algorithm.

Ethereum (Online). Available: https://www.ethereum.org.

Eyal I.; Sirer, E. G. How to Disincentivize Large Bitcoin Mining Pools. 2014 (Online). Available: http://hackingdistributed. com/2014/06/18/how-to-disincentivize-large-bitcoin-mining-pools.

Eyal, I.; Gencer, A. E.; Sirer, E. G.; Renesse, R. V. In . In 13th USENIX Symposium on Networked Systems Design and Implementation (NSDI) 16, pp. 45–59.

Greenspan, G. MultiChain Private Blockchain. 2015 (Online). Available: https://www.multichain.com/download/MultiChain-White-Paper.pdf.

Haber S.; Stornetta, W. S. How to Time-Stamp a Digital Document.   (2), 99–111.

Heimerdinger, W. L.; Weinstock, C. B. A Conceptual Framework for System Fault Tolerance. Defense Technical Information Center, Technical Report CMU/SEI-92-TR-033, 1992.

Hyperledger fabric (Online). Available: https://github.com/hyperledger/fabric.

Hyperledger iroha (Online). Available: https://github.com/hyperledger/iroha. Sumeragi (Online). Available: https://github.com/hyperledger/iroha/wiki/Sumeragi.

Hyperledger Sawtooth (Online). Available: https://sawtooth.hyperledger.org/docs/.

Intel Software Guard Extensions (Intel SGX) (Online). Available: https://software.intel.com/en-us/sgx.

Intel. Sawtooth v1.0.1. 2017 (Online). Available: https://sawtooth.hyperledger.org/docs/core/releases/latest/introduction.html.

Kiayias, A.; Russell, A.; David, B.; Oliynykov, R. In Annual International Cryptology Conference, Springer, Cham, pp. 357–388.

Kiayias, A.; Russell, A.; David, B.; Oliynykov, R. Ouroboros: A Provably Secure Proof-of-Stake Blockchain Protocol. 2016 (Online). Available: https://eprint.iacr.org/2016/889.pdf.

King, S. Primecoin: Cryptocurrency with Prime Number Proof-of-Work. 2013 (Online). Available: http://primecoin.io/bin/primecoin-paper.pdf.

King, S.; Nadal, S. PPcoin: Peer-to-Peer Crypto-Currency with Proof-of-Stake. 2012 (Online). Available: https://decred.org/research/king2012.pdf.

Lamport, L. Paxos Made Simple. (4), 18–25.

Lamport, L.; Shostak, R.; Pease, M. The Byzantine Generals Problem. (3), 382–401. Hyperledger (Online). Available: http://hyperledger.org/.

Larimer, D. Delegated Proof-of-Stake (DPOS). 2014 (Online). Available: https://bitshares.org/technology/.

Liskov, M., Fermat Primality Test. In. Springer: Boston, MA, 2005; pp. 221–221.

Liu, Z.; Tang, S.; Chow, S. S. M.; Liu, Z.; Long, Y. Forking-Free Hybrid Consensus with Generalized Proof-of-Activity.

Merkle, R. C. A Digital Signature Based on a Conventional Encryption Function. In . Springer: Heidelberg, 1987; pp. 369–378.

Miller, A.; Kosba, A.; Katz, J., Shi, E. In Proceedings of the 22nd ACM SIGSAC Conference on Computer and Communications Security, New York, NY, 2015; pp. 680–691.

Milutinovic, M.; He, W.; Wu, H.; Kanwa, M. In . Proceedings of the 1st Workshop on System Software for Trusted Execution, pp 1–6.

Nakamoto, S. Bitcoin: A Peer-to-Peer Electronic Cash System. 2008 (Online). Available: https://bitcoin.org/bitcoin.pdf.

Nxt wiki. Whitepaper: Nxt. 2016 (Online). Available: https://nxtwiki.org/wiki/Whitepaper:Nxt.

Ongaro, D.; Ousterhout, J. K. In. Proceedings of 2014 USENIX Annual Technical Conference, Philadelphia, PA, 2014; pp. 305–319.

P4Titan. Slimcoin: a Peer-to-Peer Crypto-Currency with Proof-of-Burn. 2014 (Online). Available: http://www.doc.ic.ac.uk/~ids/realdotdot/crypto_papers_etc_worth_reading/proof_of_burn/slimcoin_whitepaper.pdf. protocol. In Springer: Cham, Switzerland, 2017; pp. 357–388.

Park, S.; Pietrzak, K.; Kwon, A.; Alwen, J.; Fuchsbauer, G.; Gazi, P. Spacecoin: Cryptocurrency Based on Proofs of Space. 2017 (Online). Available: https://eprint.iacr.org/2015/528

Pommier, C. How the Private and Public Key Pair Works. 2017 (Online). Available: https://www.symantec.com/connect/blogs/ how-private-and- public-key-pair-works.

Popov, S. A Probabilistic Analysis of the Nxt Forging Algorithm., 69–83.

Quorum Chain Consensus (Online). Available: https://github.com/jpmorganchase/quorum/wiki/QuorumChain-Consensus.

Raft-based consensus for Ethereum/Quorum (Online). Available: https://github.com/jpmorganchase/quorum/blob/master/raft/doc.md.

Ren, L. Proof of Stake Velocity: Building the Social Currency of the Digital Age. 2014 (Online). Available: https://www.reddcoin.com/papers/PoSV.pdf.

Robert, E. Digital signatures 2017 (Online) Available: http:// cs.standford.edu/people/eroberts/courses/soco/projects/public-key-cryptography/dig_sig.html.

Russell A.; Zuckerman, D. In. Proceedings 39th Annual Symposium on Foundations of Computer Science, Palo Alto, CA, 1998; pp. 576–583.

Sabt, M.; Achemlal, M.; Bouabdallah, A. In . Proceedings of 2015 IEEE Trustcom/BigDataSE/ISPA, Helsinki, Finland, 2015; pp. 57–64.

Sameeh, T. Two New Models for Double Spending Attacks on Bitcoin's Blockchain. 2016 (Online).Available: https://www.deepdotweb.com/2016/12/31/two-new-models-double-spending-attacks-bitcoinsblockchain/.

Schwartz, D.; Youngs, N.; Britto, A. The Ripple Protocol Consensus Algorithm. 2014 (Online). Available: https://ripple.com/ files/ripple_ consensus_whitepaper.pdf.

Schwarz, K. Cuckoo Hashing. (Online). Available: http://web.stanford.edu/class/cs166/lectures/13/Small13.pdf.

Sompolinsky, Y.; Zohar, A. Accelerating Bitcoin's Transaction Processing: Fast Money Grows on Trees, not Chains. 2013 (Online). Available: https://eprint.iacr.org/2013/881.pdf.

Sompolinsky, Y.; Zohar, A. Secure High-Rate Transaction Processing in Bitcoin. In. Springer: Heidelberg, 2015; pp. 507–527.

Sukhwani, H.; Martinez, J. M.; Chang, X.; Trivedi, K. S.; Rindos, A. In. Proceedings of the IEEE.

Symbiont (Online). Available: https://symbiont.io/. Corda (Online). Available: https://www.corda.net/.

The Economist. The Great Chain of Being Sure about Things. 2015 (Online). Available: https://www.economist.com/news/briefing/21677228-technology-behind-bitcoin-lets-people-who-do-not-know-or-trusteach-other-build-dependable.

The Public Disputes Program. A Short Guide to Consensus Building. (Online). Available: http://web.mit.edu/publicdisputes/practice/cbh_ch1.html. 36th Symposium on Reliable Distributed Systems, Hong Kong, China, 2017; pp. 253–255.

Tromp, Cuckoo Cycle: A Memory-Hard Proof-of-ork System. 2015 (Online). Available: https://eprint.iacr.org/2014/059.pdf.

Vasin, P. Blackcoin's Proof-of-Stake Protocol v2. 2014 (Online). Available: https://blackcoin.co/blackcoinpos-protocol-v2-whitepaper.pdf.

CHAPTER 13

Controlling Blockchain Mechanism Using the Internet of Things Technology

PALVADI SRINIVAS KUMAR and THOTA SIVA RATNA SAI*

Research Scholar, Department of Computer Science & Engineering, University of Madras, Chennai, Tamil Nadu, India

Corresponding author. E-mail: thotasivaratnasai@gmail.com

ABSTRACT

Bitcoin technology has become a further revolution in information technology. Blockchain (BC) technology started its journey as bitcoin technology. Presently, the BC applications are in existence in almost all domains. This technology is utilized by private and public sectors for its advantages in terms of security, and so forth. Besides BC technology there is a technology called Internet of Things (IoT), which works by clubbing software and hardware device. The devices which are controlled by the help of IoT technology should have communication among the device and centralized server or cloud server and for storing the data. At present there are tens of thousands devices connected to Internet by help of IoT technology. We are facing several problems while synchronizing IoT devices to hardware devices. Here, we use RSA algorithm where data transmission is done by public key whereas public key is an open source and every one can use, and private key is secured one and is stored in our personal systems. We have adopted Ethereum platform for BC mechanism to easily manage, store, and maintain data. For this work we have used IoT device and we build using IoT BC.

13.1 INTRODUCTION

Internet of Things (IoT) as well as blockchain (BC) are the areas that are gaining popularity in terms of sharing and security, in the innovation hover, and in the extensive business world as well. In any case, the possibility that assembling them could bring about something much more noteworthy than the whole of its parts is something which is beginning to pick up footing. Set up them together and in principle, you have an undeniable, secure, and changeless technique for account information handled by "intelligent" devices in the IoT. BC-dependent IoT arrangements were appropriate in terms of streamlining business forms, enhancing client encounter, and accomplishing noteworthy user-friendly price. It's quoted that IoT needs BC and vice versa.

In different mechanisms IoT is able to abuse BC innovation:

1. Gaining user trust
2. Price decrease
3. Quickened information trades
4. More security

Concentrated engineering presents difficulties to anchor IoT organizations. Dealing with the huge volume of existing and anticipated information is overwhelming. Dealing with the inescapable complexities of interfacing with an apparently boundless rundown of gadgets was typical. Centralized privacy demonstrates basic for undertaking it will battle for increasing the resources in meeting the requests of the IoT.

13.2   INTERNET OF THINGS

Any article is a "thing" when it turns into an IoT if it has interfaces over the web as well as from one another. By being associated with a PC organize, the item, for example, a vehicle, and turn out to be something other than an article. Defining the term IoT is linked with hardware and software devices for successful processing and storing of the data or other items—embedded with hardware devices, programming and system network, which empower articles in gathering as well as trade information. Devices along articles identified by sensors were related to the IoT arrange, which consolidates information from the unmistakable contraptions as well as apply examination for bestow for critical data to applications attempted to address express prerequisites.

Centralized IoT architecture

It depends on incorporated, handled other mechanisms. Here devices are shared, validated, as well as linked with the cloud servers that which sport tremendous handling along capacity limits. The linking among devices happens solely over Internet with regardless of whether they are a couple of feet separated. Likewise, machine-to-machine (M2M) correspondence is troublesome on the grounds.

The unified design presents difficulties to anchor IoT arrangements. Dealing with the colossal volume of existing and anticipated information is overwhelming. Dealing with the inescapable complexities of associating with an apparently boundless rundown of gadgets is entangled. Also, the objective of transforming the storm of information into profitable activities appears to be unimaginable as a result of the numerous difficulties. The unified security display basic to endeavor that battle for increasing the resources to meet the requests for IoT.

Concept of BC

"The BC is an efficient and digitalized process of financial transfers which can be programmed for sake of recording the data for virtually everything the task we are performing on the database." (Blockchain Iot: How Will Blockchain Be integrated into the Growing Internet-of-things?, 2018).

BC implementation

BC is an appropriated database existing on various PCs in the meantime. It is continually developing as new arrangements of accounts, or "squares," are added to it. All squares are encoded uniquely, so everybody can approach all the data, however, just a client who possesses an uncommon cryptographic key can add another record to a specific chain. For whatever length of time that you remain the main individual who knows the key, nobody can control your exchanges. Moreover, cryptography is utilized to ensure synchronization of duplicates of the BC on every PC (or hub) in the system.

BC lets you demonstrate you're more than 18 years of age without disclosing your date of birth, or different subtleties. That restrains the danger of a security rupture or wholesale fraud.

From the survey overall two important properties about BC assured are:

- Transparency
- It cannot be corrupted

BC is categorized into two categories like:

1. In a public BC, here every user can read the data, write the data, and save the data.

2. In a private BC, the access permissions are given to only the authorized persons. This private BC mechanism is a costly mechanism and it is mostly used in cooperate companies, and so on.

Decentralize IoT network

Getting a regulated shared correspondence model to process the few billions of trades between contraptions will, in a general sense, decrease the costs related with presenting and keeping up gigantic brought together server cultivates and will scatter computation and limit needs over the billions of devices that shape IoT frameworks (Computer Security Division—Information Technology Laboratory—National Institute of Standards and Technology—Department of Commerce, 2018).

BC and IoT

BC development can be used in following billions of related contraptions, engage the treatment of trades and coordination between devices; consider basic venture assets to IoT industry makers. This decentralized approach would abstain from single motivations behind dissatisfaction, making a more grounded organic network for contraptions to continue running on. The cryptographic figuring used by BCs would make client data progressively private.

The record is carefully designed and can't be controlled by malevolent on-screen characters since it doesn't present in any area, and man-in-the-middle assaults can't be arranged on the grounds that there is no single string of correspondence that can be captured. BC makes trustless, shared informing conceivable and has officially demonstrated its value in the realm of monetary administrations through digital forms of money, for example, Bitcoin, giving ensured distributed installment administrations without the requirement for outsider specialists.

A champion among the most invigorating limits of the square tie is the capacity to keep up legitimately decentralized, trust record in overall trades that happen in a framework. This mechanism is essential and enables the process of every mechanism. There are a few clear favorable circumstances for building savvy machines ready to impart and work by means of the BC. Right off the bat, there is the issue of oversight. With information exchanges occurring between various systems claimed and regulated by numerous associations, a lasting, changeless record implies custodianship can be followed as information, or physical products, go between focuses in the inventory network. In the case of something turns out badly, breakages happen, information spills where it shouldn't, at that point the BC record makes it easy to recognize the powerless connection and, ideally, make a medicinal move (Open Platform for building Blockchains, 2018).

Furthermore, the utilization of encryption and circulated stockpiling implies that information can be trusted by all gatherings engaged with the production network. Machines will record, safely, subtleties of exchanges that occur between themselves, with no human oversight. Third, the "sharp contract" workplaces given by some BC frameworks, for instance, Ethereal, allow the development of understandings that will be executed when conditions are met. Fourth, BC offers the ability of amazingly upgrading the general security of the IoT condition. A critical piece of the data created by IoT is significantly personal—for point of reference, wise home devices approach close bits of knowledge concerning our lives and step-by-step plans. Over the way that it will offer new chances, it is even conceivable that BC and IoT assembly will turn into a need sooner or later. In the event that the current IoT paradigm—devices associated by means of a concentrated distributed storage and preparing service—continues, at that point frameworks are probably going to wind up progressively enlarged, as information volumes, just as the number of associated gadgets, keep on expanding.

These cloud administrations are probably going to wind up bottlenecks as the measure of information siphoned through them increments. BCs can cure this on account of their disseminated nature. As opposed to a costly, unified server farm, BC information stockpiling system is copied over the hundreds of PCs and gadgets that make up the system. This enormous measure of excess methods of information will dependably be close nearby when it's required; chopping down exchange times and significance of one server disappointment will be of no result to business movement.

13.3   ADVANTAGES OF BC IOT

Four ways IoT can exploit BC technology:

1. User trust
2. Reducing in terms of cost
3. Speed data transfer.
4. Security in flexible manner

Here mentioned are some of the methods that help in reducing the security problems in the issues like:

1. BC can be used to track the sensor data with any other malicious data.
2. Helps in the deployment mechanism in any of the IoT devices with well-suited and trustworthy manner.
3. Instead on depending on third party BC mechanism helps in reducing security problems.
4. A distributed ledger wipes out a solitary wellspring of disappointment inside the biological system, shielding an IoT gadget information from altering.
5. BC empowers gadget autonomy (smart contract), singular personality, honesty of information, and backings peer-to-peer correspondence by evacuating specialized bottlenecks and wasteful aspects.
6. The arrangement and activity expenses of IoT can be decreased through BC since there is no go-between.
7. IoT gadgets are straightforwardly addressable with BC, giving a history of associated gadgets for investigating purposes.

At the point when joined with the IoT, BC focuses on offers and new plans of action, while tending to straightforwardness, multifaceted nature, and even some security challenges encompassing information exchanges. From various perspectives, BC is the "missing connection" that empowers IoT organizations to accomplish their maximum capacity. Things being what they are, how does BC identify with IoT? In the venture, the estimation of IoT originates from the information that associated gadgets create. For instance, by breaking down the information streams from IoT-associated machines on the plant floor, producers can actualize prescient upkeep arrangements, and in this manner, increment uptime. In many use cases, such IoT information navigates crosswise over hierarchical limits and even various endeavors—giving professionals a "major picture" perspective of their tasks. The test is guaranteeing this information is precise, steady, reliable and above all, safe. Further, accommodating such information is a tedious procedure, particularly when done physically, including informational collections from unique sources that frequently don't coordinate. Here's the place BC can help. As we stated, BC encourages the consistent, perpetual, and decentralized chronicle of IoT information exchanges crosswise over gadgets and even crosswise over whole associations. How about we investigate a couple of utilization cases.

BC and IoT use cases

How about we begin with supply chains. In such mind-boggling and multilayered biological communities, BCs can ease many track-and-follow difficulties, for instance, limiting the effects of forging, item reviews, and food-borne sicknesses. Forging is a monstrous issue crosswise over general venture. In the semiconductor business alone, fakes cost US-based chip producers $7.5 billion in lost income for every year. By utilizing BC innovation, semiconductor makers can make an ethical chain of care for a chip as it navigates through each progression of the inventory network. On the off chance that a chip is accepted to be a phony, the maker can follow it through that chain of authority and evaluate where a falsifying episode happened. A similar technique can be connected to flawed chips. Thus, the sustenance business can allude to the BC to pinpoint the cause and track the location(s) of spoiled products, encouraging and notwithstanding counteracting well-being perils and destroying reviews. The sheer speed of exchange compromise is critical, permitting nourishment producers and merchants to have prompt permeability into their supply chains (IoT Security Using Blockchain, 2018).

Additionally, the sustenance business can allude to the BC to pinpoint the source and track the location(s) of corrupted products, encouraging and notwithstanding avoiding well-being perils and decimating reviews. The sheer speed of exchange compromise is urgent, permitting sustenance makers and distributors to have quick perceivability into their supply chains. Walmart showed such capacity by following a bundle of mangos back to the ranch from which it began. With BC and IoT, the procedure took simply 2.2 seconds, though it would have taken almost seven days utilizing customary computerized and paper-based strategies. Be that as it may, BC and IoT likewise hold incredible guarantee for making new incentives. Take accommodating well-being records for instance. With a moment, the precise and secure perspective of a patient's restorative history, specialists can guarantee they don't recommend a drug that could antagonistically affect the patient, in light of past or present medicinal conditions. This utilization case could prompt more prominent patient security and enhanced well-being results. BC could likewise help reinforce information security, decentralizing individual information put away in different storehouses, and giving clients understanding into who has gotten to it. With respect to new plans of action, BC can empower smaller scale installments, enabling organizations to bring down their exchange charges, cut down exchange times, and make a straightforward and reasonable installment conveyance framework. A model where micropayments are utilized is in the music-spilling industry.

13.4   WHAT'S NEXT FOR BC?

We've just touched the most superficial layer of BC's capacities, particularly when it is matched with IoT—and the future looks splendid. Indeed, the worldwide BC innovation advertise is relied upon to reach $7.59 billion by 2024, at a 37.2% CAGR. Be that as it may, a dominant part of the present BC endeavors is still in innovation disclosure and evidence of-idea stages.

To meet the undertaking prerequisites, we will require the business to create arrangements dependent on builtup systems that will permit interconnectivity between divergent BC systems. Such arrangements ought to be very versatile and give sending adaptability, fine-grained data classification, and solid security with prescient risk examination and agreement control. What's more, obviously, they have to convey genuine execution fitting for the utilization case. Meanwhile, I would urge IoT professionals to investigate how BC can apply to their information streams and explicit use cases to decide if BC is the correct device to tackle their issues—and whether it is really the missing connection in their IoT organization.

As the number of sensors in vehicles, production line apparatus, structures and city framework develop, organizations are searching for a protected and computerized method for empowering a work to arrange for value-based procedures. BC seems to best fit that bill. A significant bit of the evaluated 46 billion modern and venture gadgets associated in 2023 will depend tense figuring, Juniper stated, so tending to key difficulties around institutionalization and sending will be critical. BC, the electronic, distributed ledger technology (DLT) that additionally contains a business robotization programming segment—known as self-executing "keen contracts"—could offer an institutionalized strategy for quickening information trade and empowering forms between IoT gadgets by evacuating the go-between. That mediator: A server that goes about as the focal correspondence represented solicitations and other traffic among IoT gadgets on a system.

13.5   IMPROVING A ROBUST SUPPLY CHAIN

In a customary supply chain, Milosevic clarified, a focal server validates the development of products and materials starting with one area then onto the next. Or instead, a focal expert could choose to end certain procedures dependent on predecided standards. In a circulated BC IoT arrange, the IoT gadgets on a distributed, work system could confirm exchanges and execute exchanges dependent on predecided tenets—without a focal server.

BC records decline the time required to finish IoT gadget data trade and preparing time. The ascent of edge figuring is basic in scaling up tech arrangements, attributable to diminished transmission capacity necessities, quicker application reaction times, and enhancements in information security, as per Juniper inquire about. BC specialists from IEEE trust that when BC and IoT are joined they can really change vertical enterprises. While money-related administrations and insurance agencies are as of now at the cutting edge of BC improvement and sending, transportation, government, and utilities parts are currently captivating progressively because of the substantial spotlight on process proficiency, supply chain and logistics, said David Furlonger, a Gartner VP and research individual. The procedure for following medication shipments, be that as it may, is profoundly divided. Numerous pharmaceutical organizations pay store network aggregators to gather the information along the way to meet the administrative models. One misguided judgment about BC is that it replaces heritage frameworks (BIoT: Blockchain Based IoT – Coinmonks – Medium-Sarang Chaturvedi, 2018).

A year ago, SAP banded together with IBM to demonstrate how IoT and BC could automate a pharmaceutical store network for both following and announcing purposes. SAP consolidated its Leonardo IoT programming framework with IBM's BC cloud administration to make a working model of a framework that could follow and oversee pharma supply chains utilizing smart contract rules. One misinterpretation about BC, is that it replaces traditional frameworks. "Truth be told, BC is a layer over big business applications." SAP likewise as of late finished two proof-of-concept (Poc) BC organizations with clients: one was a littler production network test assessing a large number of exchanges on a Hyper ledger BC utilizing shrewd contract innovation in IoT gadgets; the second was a lot bigger one that spoke to billions of exchanges among 15 distinct clients utilizing Multichip open-source BC programming without smart contract code on the gadgets. The littler Poc with smart contract innovation functioned admirably, yet was progressively costly to set up in light of the fact that it required a BC designer—somebody that is hard to find—to compose the code, said Gil Perez, head of Digital Customer Initiatives for SAP.

"In the little pilot, the expense of tasks and overhead was fairly high. Along these lines, practically it surpassed desires, yet from money related point of view it was exceptionally testing," Perez said. The second Poc tended to both versatility and cost, in that it demonstrated the BC could scale to big business levels. The MultiChain PoC was not as productive as the hyper ledger show, but rather it cost less and it met the business necessities,

Perez said. "The way that you put the rationale on a server doesn't mean it's not computerized," Perez stated, alluding to the bigger Poc. SAP has been working with around 65 of its clients to create BC expanded programing—something Perez said will before long be accessible. At its Sapphire Conference in May, SAP declared a BC cloud benefit for its clients—something that is likewise turning into a mainstream alternative for organizations that would prefer not to consume capital so as to test circulated record innovation (McGrath, 2019).

IBM likewise as of late propelled an IoT-to-BC benefit as an extra to its current IoT Connection Service. It empowers IoT gadgets, for example, RFID area chips, scanner tag sweeps, or gadget-detailed information, to be transmitted to a permission BC on IBM's cloud benefit. That BC-based system would then be able to be utilized by a business system of PCs to approve provenance. Gadgets ready to convey information to BC records can refresh or approve shrewd contracts. For instance, as an IoT-associated shipment of merchandise moves along various circulation focuses, the bundle area and temperature data could be refreshed on a BC according to IBM. New uses for BC will continue grow because of the DLT's ability to provide new forms of security, as per B2B reviews platform G2 Crowd. "Data security and corporate uprightness have both taken a blow after occasions like the Equifax disaster, and organizations are putting resources into BC as a precautionary measure," G2 Crowd said in a report.

BC empowers secure access to IoT sensors

For instance, worldwide wireless system innovation supplier ABB Wireless embraced BC as a strategy for conveying decentralized security administrations for modern frameworks in enterprises, for example, utilities, oil and gas, and transportation. On a framework with thousands or a huge number of IoT hubs, the likelihood of hacking the system is remote, in the best case scenario. "On the off chance that you include a million progressively keen meters to a remote system, you've recently made that arrange more enthusiastically to hack. Though in a conventional system, the more units you include, the more introduction there is to hacking," said Age CEO Duncan Great wood.

ABB Wireless is using a BC platform developed by Xage, a startup that officially launched earlier this month. The BC application contains an encoded and permanent table of security qualifications, which permits field laborers to sign into a gadget—regardless of whether the substation is separated from a utilities' focal server farm because of a mishap, for example, a rapidly spreading fire. "Everybody's terrified to death that somebody will gain power of the matrix," said Paul Gordon, VP of designing and tasks at ABB Wireless. ABB Wireless utilized Xage's security application running nervous doors inside various segments at power utility substations. The mesh arrange empowers secure, remote access to IoT gadgets to control the substations, taking into account everything from review support information to rerouting power. A lot of research is going on this domains to make the things smart and more case sensitive. "This gives an answer scalable, so security doesn't turn into a gigantic weight. It takes into consideration an increasingly adaptable arrangement while addressing the requirements of a profoundly anchored condition." Consolidating the unchanging nature of a BC-circulated record with encryption implies that the more end hubs that are included; the more secure the system progresses toward becoming, not at all like customary social database frameworks that have a solitary purpose of access. BC-on-IoT gadgets are increasingly secure in light of the fact that a digital aggressor would need to break into a larger part of the hubs to increase controllable access to a framework.

The BC is a disseminated database innovation that gives hard to alter, record records. It permits stockpiling of all exchanges into permanent records and each record conveyed crosswise over numerous member nodes. The security originates from the utilization of solid open key cryptography, solid cryptographic hash, and complete decentralization. Blocks are the key idea of innovation. They are little arrangements of exchanges that include occurred inside the framework. Each new block stores reference of the past exchange by including a SHA-256 hash of the past exchange. Along these lines, it makes a "chain" of blocks and subsequently the name. Squares are computationally hard to make and takes different specific processors and critical measures of time to produce. Since producing a block is troublesome and to alter one block, one needs to alter the past block and afterward needs to pursue to transform it totally, BC innovation is viewed as alter safe.

13.6 HOW IOT CAN LEVERAGE BC TECHNOLOGY

With IoT beginning to get into the standard business, the key difficulties of the innovation are quick rising. One of the key regions of IoT arrangement is security. Following are the key security challenges for IoT foundation and administrations:

- With the possibility of gadgets in the framework developing exponentially, it is an immense test to distinguish, validate, and secure the gadgets (Amit Goel, 2018).
- A brought together security model will be extremely troublesome and costly proportional, keep up and oversee.
- A concentrated security foundation will present a solitary purpose of disappointment and will be an obvious objective for DDoS assault.
- Centralized foundation will be hard to execute in modern setup where the edge hubs are across the board geologically.

BC innovation is by all accounts a suitable option because of the key qualities depicted previously.

Cases where BC can be used

It very well may be utilized to make an anchored work organize that will permit IoT gadgets to interface safely and dependably keeping away from the dangers of gadget parodying and pantomime. Each IoT node can be enlisted in the BC and will have a BC id, which will particularly recognize a gadget in the all-inclusive namespace. For a gadget to interface another gadget, one will utilize the BC id as URL and will utilize its neighborhood BC wallet to raise a character ask. The wallet will make a carefully marked demand and send to the objective gadget that will utilize BC administrations to approve the mark utilizing general society key of the sender. Along these lines, M2M validation can occur without the need of any incorporated referee or administration. For a gadget that is compelled by an asset can be associated with intermediaries where the wallet can be put away. This will present some type of conglomeration yet it will be genuinely restricted. The above conceivable arrangement will be material to a wide scope of IoT administrations. A portion of the models will be keen social insurance-associated vehicles, coordination, transportation, and so forth.

Cases where BC is not the best solution

One key advantage of utilizing BC innovation is its utilization as a circulated account framework. It permits to safely compose changeless records. To do that, it utilized solid cryptography and replication. For instance, in supply change the board, a committal needs to go by means of a progression of exercises and the status of the bit of a thing can be checked by means of RFID and recorded utilizing BC innovation. A slight spontaneous creation may make BC adjusted to close time circumstances. A presentation of total storing hub at the nearest separation of the sources can be utilized as a dealer among source and BC administrations. Be that as it may, this will be a deviation from the key quality of BC and must be utilized after cautious thought.

13.7 FUTURE CHALLENGES

With quantum PC getting to be reality, the dread is that it would break open key encryption. Driving associations on the planet for example NIST has begun activities to create post-quantum cryptography. All in all, that brings up the issue: how safe will BC innovation stay in future? Hard to reply. I will rather introduce some important information. D-Wave framework declared the accessibility of 2000 quit framework. Presently, to factor 1024-piece RSA key, it takes 2000 bits and dominant part utilizes RSA key size bigger than that. Given, that D-Wave innovation is debated by specialists and the scaling quantum PC isn't like a typical PC, it isn't deterministic to what extent before a substantial RSA key (> 1024 bits) can be animal constrained.

In ongoing investigations and discussions, specialists have recognized a few difficulties in accomplishing this future, which could be limited to the accompanying: (Banking Giants Embracing Bitcoin and Blockchain Tech-Yessi Perez, 2018).

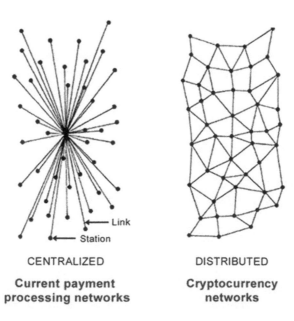

CENTRALIZED

Current payment processing networks

DISTRIBUTED

Cryptocurrency networks

- **Scalability** – Notwithstanding the developing number of seas/cloud benefits, the present Internet of billions of Things won't scale to the Internet of several billions of Things due to moderate selection, costs and the multifaceted nature of most IoT arrangements, reports the IBM Institute of Business Value .
- **Lack of standardization and protocolization** – Removing from the image size and structure, up until this point, no exertion has been put into making a solitary correspondence standard and there is no rise of a worldwide programming arrangement.
- The non-presence of an ideal cloud/web benefit with 100% uptime— on the off chance that we think back, each major seas/cloud benefit has had its issues, from bonnets attacking Cloud Flare to big outages causing Skype calls services to go down , we've seen it all.

What could be the solution to these difficulties?

Most, if not these difficulties in growing the IoT can be tended to by innovative leaps forward, for example, appropriated accord in software engineering, progressions in cryptography, and headways in shared system advances (IBM Blockchain – Enterprise Blockchain Solutions and Services, 2018). Enter the BC. It's an unchanging record of computerized occasions (exchange of significant worth or potentially state) shared distributed between various gatherings. It must be refreshed by agreement of a lion's share of the members in the framework and, once entered, data can never be eradicated. It is continually developing, as mineworkers add new squares of information to it (at regular intervals or like clockwork, depending of the BC) to record the latest occasions. The squares are included a direct, sequential request and each full hub (for example processing gadget associated with the system utilizing the digging customer for approval) has a duplicate of the BC, which is downloaded naturally after joining. The approval procedure (or, in BC wording, the mining procedure) is by one way or another troublesome, as in it needs a great deal of vitality as proof-of-work (Pow), in this way, for an awful on-screen character to upset the system, it requires somewhere around somewhat more than the whole registering intensity of the system, making it unfeasible to decimate.

The BC innovation—recently utilized in advanced monetary forms, for example, Bitcoin—has surely caused a tempest in the realm of back and keeping money (or, all the more essentially, FitnTech). Inventive new organizations are starting at now applying this development to chop down the expense of exchanging cash . Huge venture funds , close by significant financial monsters are putting many millions in such leap forward activities.

The ideal case of how this is changing the financial business is the declaration that 22 noteworthy banks are joining a consortium to shape a typical structure for utilizing BC innovation (the R3 BC consortium). As BC technology has started with many problems out of which some of the problems were solved. They are:

- It is genuinely versatile, as it can scale a similar way a deluge can to a huge number of companions;
- It can possibly institutionalize the whole system.
- The security is inherent: the way that the BC is conveyed crosswise over a huge number of PCs implies that hacking it is practically inconceivable or unfeasible (truth be told, this is one of the primary reasons why we see such a major promotion in the saving money industry over the uses of this innovation).
- It is straightforward, as the cryptographic open location of the sender and collector of each occasion is recorded, and everything is accessible for assessment.
- It can likewise offer extraordinary enhancement as far as security, as clients are pseudonymous and can perform exchanges in a flash without the requirement for any close to home verification (IBM IoT Blockchain Service, 2018).
- It has no downtime and it's resistant to oversight as no single expert has general power over it (because of its deconcentrated structure). On the off chance that the database of an establishment or a cloud framework goes down, clients will be not able to perform information exchanges with it, which, much of the time, can have genuine results. With BC, clients can have constant, open access, with no potential danger of intrusion.

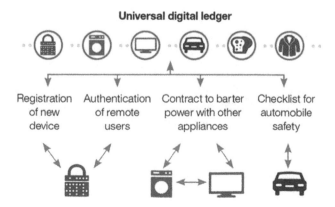

Universal digital ledger

Registration of new device | Authentication of remote users | Contract to barter power with other appliances | Checklist for automobile safety

Another worldview in structuring IoT-empowered applications can rise up out of the further developed brilliant BC advances that are being created. Maybe this space also will be changed, as on account of the Fitch condition. There are many undertakings/associations that guarantee to do only that. Among them, the most eminent are the accompanying:

- Ethereum – Ethereal is an amazingly unpredictable task. It centers on the improvement time and security of little applications, and the capacity of various applications to interface productively.
- Eris Industries – Eris is a free programming platform that enables anybody to manufacture their own protected, minimal effort, run-anyplace information framework utilizing BC and brilliant contract innovation.
- Expanse EXP – Expanse is decentralized cryptographic data, appli-cation, and contract stage. It is among the first of such to be genuinely appropriated, fairly controlled, and network oversaw. Using shrewd contracts and decentralized BC innovation, it is run not by any one individual or gathering, but rather by the clients of Expanse itself.

All in all, regardless of whether the IoT and BC arrangement will prevail upon the IoT and Cloud arrangement is up in the air. Surely, the few new businesses that are as of now testing this field are insufficient to make the system impact required for most organizations to participate in utilizing a similar thought, however, history is brimming with capricious occasions. For sure, the Internet itself could be noted as one of them. We presently have become used to this universal innovation and will in general overlook its modest beginnings, yet the exact medium that we are utilizing to peruse this article is a case of system impact that has totally changed the media communications industry. The BC is without a doubt promising and it is additionally certainly worth specialists' consideration – a specific subject meriting thought being the best equipment choices for installing it in various IoT applications (IoT Security Using Blockchain, 2018).

It has turned into a progressive innovation as it has no physical shape and has to arrange based limitations. All the applications and locales about monetary exchanges have turned out to be completely secure because of this innovation. Because of the gigantic capability of the BC, tech-specialists are discovering approaches to fuse it with different strategies and in various parts like IoT. The IoT causes gadgets to drive information to the records of BC for including the common exchanges with change evidence records.

With the assistance of IoT, the system clients can confirm their exchanges. Each member is in charge of their own exchanges and, subsequently, no debate can emerge because of any obstruction. Virtual exchanging and following any important thing without a particular control point can help in making a cost-effective business system, and this is conceivable because of the conveyed record innovation. BC and IoT work superbly together and they have mixed well after some time. IoT chips away at a brought together model, that is, a customer demonstrates as far as business. This mix furnishes an entryway to speak with physical items interconnected by means of the web. These gadgets can be kept running from any area. These gadgets keep running on cloud servers with a boundless limit of capacity and IoT can help in confirming and distinguishing those gadgets. The total computerized foundation is restored by the mix of IoT and BC. Money-related exchanges happen effectively in this space. Utilizing IoT and BC together, the network and security of the gadgets have enhanced in a superior and effective way. In addition, BC is popular and utilized in practically all organizations for meeting the necessities and building up a superior framework. Indeed, even Walmart utilizes BC innovation for boosting proficiency and precision in the work process.

13.8   WORKING MECHANISIM OF BC AND IOT?

Presently, you realize that a decentralized arrangement of BC can store boundless records of each money-related exchange. Thus, having a decentralized framework implies that the incomparable individual approaches the all-out budgetary subtleties. Each machine has distinctive data spared in squares on an allowed system. Each square contains a hub that spares data sequentially. There is nothing to stress over a rupture in a domain because of high-security association got from the mix of IoT and BC.
Benefits of using IoT and BC together:

- No need of records updating
- Peer-to-peer connection
- Uses algorithms
- Maintains transparency

Each exchange has a nitty-gritty, further exchange identified with it. The individual who approaches it can see the whole subtleties for each. For confirmation of individual character, BC innovation gives open and private alphanumeric key for encryption.

More about the IoT and BC combination

IoT with BC can help with following understandings since organizations can follow the historical backdrop of a procedure or hardware in an ineradicable record. Likewise, this record can be imparted to related safety net providers or offices.

The engineering of IoT has a structure that can hold up under a few security dangers. It can additionally be useful in dropping the powerlessness dimension of BC organize, which can happen because of adjustment assaults, the 51% assault, and DDoS assault. Thus, IoT helps in giving BC a structure that is secure, lightweight, and private alongside the points of interest that BC innovation as of now offers, bringing out upgraded security that can keep online dangers under control and ensure us amid each exchange .

BC innovation could demonstrate particularly essential as the IoT keeps on multiplying. A developing number of gadgets are presently being associated with the Internet, taking into consideration remote control, expanded information gathering, and different exercises. In any case, really dealing with these IoT gadgets and the information they produce is a test. BC could give an answer. Numerous business chiefs in the fund, drug, and different ventures have just referred to BC as a potential distinct advantage. So, what is BC at any rate, and what does it have to do with IoT? It serves to initially see a few essentials about bitcoin (Security Trends, 2018).

13.9   BITCOIN AND THE BC

Bitcoin hails itself as distributed digital money. When we state "shared," we truly say that bitcoin is decentralized and not controlled by any administration or specialist. Rather, the network itself deals with the money. National cash is upheld by the governments and supply is controlled by a national bank. Such a bank can either decrease or expand the cash supply, in this way influencing the general estimation of the money itself. National banks can likewise set financing costs and settle on different choices that impact sly effect on national monetary forms. With bitcoin, there is no focal specialist. Rather, the network all in all controls the supply of bitcoin, screens exchanges, and generally keeps up the cash itself. The BC assumes an essential job in this network-driven administration. Fundamentally, there is an "open record" that tracks each bitcoin exchange at any point led. As you can envision, this record is very enormous and is additionally always developing. The BC keeps up this spending record by enabling the network to grow and affirm increases of "hinders" to the chain. "Excavators" mine bitcoin squares, which are squares of exchanges. In return, these mineworkers are remunerated with recently printed bitcoins. This implies the diggers are driven by motivations to mine new bitcoins. Each exchange is affirmed by the mineworkers, and just when exchanges are checked, they are added to the record. While general society record is mysterious, it's additionally "open." Anyone can perceive any exchange, in spite of the fact that the personalities of the general population completing the exchanges are covered up.

BC IoT

The BC is both basic and splendid. Consider what number of exchanges is currently happening with bitcoin and that it is so hard to track and process those exchanges. Truth be told, more than 400,000 exchanges are at present led every day. It would take an expansive bank, credit handling organization, or focal fiscal expert to process and track these exchanges. However, there is no brought together specialist like this in bitcoin, and no unified preparing focus either. In any case, the network can process and report about a half million exchanges day by day. The BC takes into account appropriated handling by remunerating mineworkers for conveying their preparing capacity to hold up under. Dispersed preparing could demonstrate imperative, and with the BC, it's likewise effective. As the IoT develops, more information will pour in from more sources. With IoT, handling force will likewise be circulated crosswise over incalculable gadgets. Now and again, preparing force may go unused. At different occasions, you may end up sufficiently lacking handling capacity to do what you need.

Consider your PC, cell phone, and other shrewd gadgets. When you're not effectively utilizing these gadgets, most if not the majority of their handling power is sitting unused. It turns into a squandered asset when not put to utilize. However, on the off chance that you wind up attempting to accomplish something handling concentrated, state gathering code for a product program you're building, you may find that you do not have the assets to take care of business. With the BC innovation, it's less demanding to convey handling capacity to where it is required. Returning to bitcoin, there are really mining societies that individuals can join with their PCs. When you go along with one of these mining gatherings, they can utilize your PC's preparing power in a joint effort with different PCs to mine bitcoin. So as opposed to closing down your PC for the night, you may credit it to a mining society so they can mine bitcoin. For organizations, such adaptable preparing could come in particularly convenient. With the IoT, the BC innovation could be given something to do following and preparing the gigantic stream of information that will pour in. A BC could be set up with a record for the information. When handling power is required, those gatherings that help the BC, conceivably importance end-clients, could coordinate their preparing power toward keeping up the BC in return for some sort of remuneration. Numerous IoT gadgets are being outfitted with processors. Indeed, even little gadgets, similar to surveillance cameras, are being furnished with installed chips. Be that as it may, onload up preparing abilities may not be sufficient to deal with each need, constantly. With BC, IoT preparing force could possibly be redistributed when required (Chancellor on Developing Fintech, 2018). BC could be huge for IoT security

The Internet has been an astonishing and amazing asset for sharing data, spreading learning, and generally empowering quick, solid correspondence. Be that as it may, the Internet accompanies its very own dangers, and particularly security concerns—email tricks, infections, hacked accounts, and so on. As the IoT develops, so too will the security dangers. Before long, a large number of our gadgets will be empowered with sensors, for example, cameras and amplifiers. Imagine a scenario where deceitful gatherings could hack these gadgets to keep an eye on you, take individual data, or generally hurt you. This makes security is a noteworthy test for IoT. Numerous security specialists trust that BC can be utilized as a security arrangement. BCs are great for following and confirming information. Outfitted with different encryption techniques and confirmation keys, BCs can be made monstrously secure. Dispersing preparing and multistep check makes BCs hard to hack. So, could BC strategies be given something to do to confirm characters and decide control? Security specialists accept so. As the IoT develops, checking information, exchanges, access, control, and different viewpoints will end up imperative. For organizations moving IoT gadgets, that may mean setting up tremendous servers and propelled verification and security frameworks. Nonetheless, doing as such can get costly and cumbersome. Further, the more information is assembled in one place, the higher the dangers of security rupture. Consider the gigantic Equifax break and different infamous cases. The answer for this is set up BC frameworks that can assemble information, check access, and process exchanges, in addition to other things. Security specialists are simply beginning to investigate the maximum capacity for BC. Up until this point, be that as it may, the Internet security industry is peppy, with many idea pioneers thinking about whether BC is the following enormous thing. There are dangers and difficulties, however. For instance, BC works to some degree like a casting a ballot system. A group of PCs process data, looks at, and afterward orders the information. In principle, if a programmer can pick up control of at any rate 51% of the handling power, they could basically capture the whole BC. With bitcoin, for instance, in the event that somebody anchored 51%, they could essentially take all the bitcoins. The bitcoin network comprehends this test by disseminating preparing capacity to and a whole lot bigger number of clients. It would be troublesome, close outlandish indeed, for programmers to anchor 51% of the preparing power that at present runs bitcoin. Be that as it may, shouldn't something be said about littler systems and BCs? Suppose, a network sets up a BC to process different keeping money exchanges. The BC is then disseminated among clients. On the off chance that there are just a couple of hundred individuals preparing the BC, programmers may be able to pick up power over it. Security specialists are at present working out the arrangements, yet the danger of 51% assaults stays high and genuine. BC IoT applications are as a rule completely investigated seconds ago. The BC is as yet a moderately new technique and real organizations are simply beginning to comprehend and put resources into the BC. Significant organizations, including Google, Goldman Sachs, and different funding firms, are as of now putting intensely in the BC innovation. Correct BC applications for IoT security and dispersed preparing are being worked out. However, the development of the IoT and BC's capacity to deal with tremendous measures of information and to adaptable allot handling capacity to where it's required makes the two ideas an ideal fit for one another. Later on, BC IoT applications might be both normal and progressive (Open Platform For building Blockchains, 2019).

13.10   IOT CHALLENGES IN HEALTHCARE DOMAIN

Driving human services associations are utilizing the amusement changing intensity of the IoT and BC to enhance persistent results and streamline interior tasks. Separately, and all the more essentially when joined, IoT and BC — particularly when matched with man-made brainpower (AI) and machine learning—give considerably more conceivable outcomes to human services. The development dimension of IoT is well in front of the BC. Varieties of the IoT show—utilizing various information gathering sensors to assemble continuous data—have been utilized for quite a long time in the assembling and utility enterprises. What's going on for human services is the term Internet of Healthcare Things (IoHT), which alludes to an associated framework of gadgets and programming applications that can speak with different social insurance IT frameworks. Numerous social insurance associations are as of now utilizing IoT, from checking babies to following stock and looking after resources. A report by the executives counseling firm Frost and Sullivan predicts that the quantity of IoT gadgets will ascend from about 4.5 billion out of 2015 to upward of 30 billion by 2020.

Places where IoT is used in healthcare domain?

The suitable reaction is that there are two unquestionable orders of usage cases—one for clinical organizations and the other for help exercises. In clinical settings, IoT is upgrading industrious-driven activities with remote patient monitoring (RPM). IoT also pushes clinical fundamentals by about after key signs and some different pointers basic to the examinations, for instance, glucose levels and weight designs. IoT benefits support undertakings with apparatus-driven sensors and data-gathering capacities that can lessen costs through better utilization of flexible remedial assets. Sensors give the workplaces staff nonstop information about the use rates and region of mechanized X-bar equipment, ventilators, and other versatile resources with the objective that they can be doled out more sufficiently and quickly discovered when required, saving thought experts imperative time. Additionally, IoT material data sources can exhibit specialists the steady execution status of expensive machines, for instance, MRI equips. Social protection pioneers are in like manner evaluating the mix of IoT and augmented reality (AR) advancement to make automated twins. Keen and exceptionally visual AR interfaces may cautiously replicate complex mending focus apparatus to give experts and clinicians down-to-earth, hands-on getting ready openings

BC source of truth

More current than IoT, BC is picking up consideration on account of its capacity to create and safely disperse changeless, unalterable records of exchanges. BC makes successions of exchanges, known as squares, and records them in a progressing chain of occasions that can be shared among individuals from a system. Since the squares are ensured utilizing progressed cryptographic innovation, the records are essentially difficult to change, as indicated by BC advocates. Human services experts are paying heed. Research by the administration counseling firm Deloitte discovered that 35% of administrators at well-being and life sciences organizations are wanting to execute BC inside the following a year.

13.11   WHERE DOES BC FIT IN HEALTHCARE?

BC has expansive ramifications for the business, including the capacity to disentangle and enhance security and exactness for awkward, wasteful procedures. Models incorporate streamlining claims arbitration, quicker restorative protection enrolment, and enlarging B2B movement over the social insurance esteem chain. At the point when BC is joined with IoT, secure, unalterable BCs can likewise diminish chances by graphing the chain of ownership of medicinal gadgets and pharmaceuticals, enabling items to be followed through the production network. This information gives a significant legal trail if quality issues emerge after conveyance of a thing. Thus, if a maker recognizes an issue with a gadget or pharmaceutical, the BC can enable merchants to speed up reviews by rapidly deciding the area of stock over the production network to keep it unavailable for general use. Other BC openings incorporate quicker and the sky is the limit from there proficient credentialing of workers, on account of the capacity to confirm unalterable records of parental figures. Medicinal services administrators arranging IoT and BC executions ought to consider two supplemental advancements that can speed reception and time to esteem (Security Trends, 2018). The first is the utilization of AI, which totals and laterextricates experiences by perceiving examples and relationships crosswise over extensive volumes of information. Half-and-half mists are the second key to a solid IoT and BC establishment. These new crossbreed cloud-at-client models give very versatile open cloud applications and administrations while keeping clinical information behind big business firewalls to adjust to hierarchical and administrative necessities.

Time to act

The most recent computerized advancements are making open doors for reshaping how medicinal services associations convey quiet administrations, enhance results, upgrade clinician fulfillment and oversee costs all the more viable. Be that as it may, doing as such requires an unmistakable comprehension of what IoT and BC can convey in the prompt future, and as they keep on developing after some time. To do that, officials ought to evaluate the present condition of these advancements and how they can be joined with the assistance of cloud administrations for considerably more noteworthy effect.

13.12 CONCLUSION

BC and IoT combination will turn into a need eventually. On the off chance that the current IoT paradigm—devices associated by means of unified distributed storage and handling service—continues, at that point frameworks are probably going to end up progressively enlarged, as information volumes, just as the number of associated gadgets, keep on expanding. These cloud administrations are probably going to end up bottlenecks as the measure of information siphoned through them increments. BCs can cure this on account of their appropriated nature. Subsequently, the merger of these two inclining advances is inescapable. These two advancements supplement just as help one another. It is notwithstanding being said that IoT needs BC and BC needs IoT.

KEYWORDS

- **internet**
- **server data**
- **storage**
- **devices**
- **connection oriented**
- **connection less**

REFERENCES

Blockchain in the Real World: 3 Enterprise Use Cases-Lucas Mearian(n.d.).https://www.computerworld.com/article/3243700/emerging-technology/blockchain-in-the-real-world-3-enterprise-use-cases.html?upd=1529435637731 (accessed Feb 11, 2019).

Money Transfer Companies Using Blockchain Technology-Amit Goel (n.d.).http://letstalkpayments.com/11-money-transfer-companies-using-blockchain-technology-2/ (accessed Feb 15, 2019).

5 Things Supply Chain Leaders Should Know About Blockchain-Christy Pettey(n.d.).https://www.gartner.com/smarterwithgartner/5-things-supply-chain-leaders-should-know-about-blockchain/ (accessed Feb 10, 2019).

Banking Giants Embracing Bitcoin and Blockchain Tech-Yessi Perez (n.d.).http://www.coindesk.com/8-banking-giants-bitcoin-blockchain/ (accessed Feb 19, 2019).

A Guide To the Internet Of Things Infographic(n.d.).http://www.intel.com/content/www/us/en/internet-of-things/infographics/guide-to-iot.html (accessed Feb 13, 2019).

Australian Hospital + Healthcare Bulletin: Industry News, Comment, Feature Articles, Case Studies and New Products (n.d.).https://www.hospitalhealth.com.au/content/technology/article/IoT-and-blockchain-unique-opportunities-for-healthcare-1007804102#axzz5cUOouU2D (accessed Feb 20, 2019).

BIoT: Blockchain Based IoT – Coinmonks – Medium-Sarang Chaturvedi. (n.d.).https://medium.com/coinmonks/bIoT-blockchain-based-IoT-bd913162b6d1 (accessed Feb 10, 2019).

Blockchain IoT: How Will Blockchain Be Integrated into the Growing Internet-of-things? (n.d.).https://businesstown.com/blockchain-IoT-be-integrated-into-the-growing-internet-of-things/ (accessed Feb 20, 2019).

Chancellor on Developing Fintech (n.d.).https://www.gov.uk/government/speeches/chancellor-on-developing-fintech (accessed Feb 15, 2019).

Computer Security Division- Information Technology Laboratory- National Institute of Standards and Technology - Department of Commerce(n.d.).http://csrc.nist.gov/groups/ST/post-quantum-crypto/ (accessed Feb 12, 2019).

D-wave Systems(n.d.).https://www.dwavesys.com/press-releases/d-wave-systems-previews-2000-qubit-quantum-system (accessed Feb 13, 2019).

Ethereum Project(n.d.).https://ethereum.org/ (accessed Feb 17, 2019).

Expanse Tech(n.d.).http://www.expanse.tech/ (accessed Feb 18, 2019).

Finance Heavyweights Invest in Blockchain Start-up-Nicole Bullock (n.d.).http://www.ft.com/cms/s/0/dc1d46c2-572d-11e5-9846-de406ccb37f2.html (accessed Feb 16, 2019).

How Using Blockchain in the Supply Chain Could Democratize Innovation Itself-Conner Forrest(n.d.).https://www.zdnet.com/article/how-using-blockchain-in-the-supply-chain-could-democratize-innovation-itself/ (accessed Feb 10, 2019).

https://erisindustries.com/ (accessed Feb 18, 2019).

https://www.ibm.com/support/knowledgecenter/en/SSQP8H/iot-connected-products/overview/overview.html (accessed Feb 11, 2019).

https://www.rs-online.com/designspark/when-the-blockchain-technology-meets-the-internet-of-things (accessed Feb 19, 2019).

https://www.sap.com/documents/2016/12/84e838d6-9d7c-0010-82c7-eda71af511fa.html (accessed Feb 10, 2019).

IBM Blockchain - Enterprise Blockchain Solutions and Services(n.d.). https://www.ibm.com/blockchain (accessed Feb 11, 2019).

IBM IoT Blockchain Service (n.d.).https://www.ibm.com/us-en/marketplace/iot-blockchain/details(accessed Feb 11, 2019).

IoT and Blockchain: A Fruitful Combination? - Dzone Iot-Bhushan Aher (n.d.).https://dzone.com/articles/how-the-combination-of-IoT-and-blockchain-is-fruit (accessed Feb 20, 2019).

IoT Could Be the Killer App For Blockchain-Lucas Mearian (n.d.).https://www.computerworld.com/article/3284024/blockchain/IoT-could-be-the-killer-app-for-blockchain.html (accessed Feb 12, 2019).

IoT Security Using Blockchain(n.d.).https://www.uk.sogeti.com/content-hub/blog/IoT-security-using-blockchain/ (accessed Feb 13, 2019).

Mobile Ad Networks as DDoS Vectors: A Case Study-Marek Majkowski (n.d.).https://blog.cloudflare.com/mobile-ad-networks-as-ddos-vectors/ (accessed Feb 14, 2019).

Open Platform For building Blockchains (n.d.).https://www.multichain.com/.

R3.com(n.d.).http://r3cev.com/ (accessed Feb 17, 2019).

Security Trends 2018: Blockchain Explained(n.d.).https://blog.g2crowd.com/blog/trends/cybersecurity/2018-cs/blockchain/ (accessed Feb 11, 2019).

Semiconductor Counterfeiting Is a Global Problem-Dylan McGrath. (n.d.).https://www.eetimes.com/author.asp?section_id=40&doc_id=1332064 (accessed Feb 10, 2019).

Skype Outage Sees Internet Calls Go Down in Many Countries-James Titcomb (n.d.).http://www.telegraph.co.uk/technology/news/11879664/Skype-outage-sees-internet-calls-go-down-in-many-countries.html (accessed Feb 14, 2019).

Technology in the Blockchain Market Is Expected To Grow-Global Banking & Finance Review(n.d.).https://www.globalbankingandfinance.com/technology-in-the-blockchain-market-is-expected-to-grow/ (accessed Feb 10, 2019).

Thirteen More Top Banks Join R3 Blockchain Consortium-Jemima Kelly (n.d.).http://www.reuters.com/article/2015/09/29/banks-blockchain-idUSL5N11Z2QE20150929 (accessed Feb 17, 2019).

What Blockchain Means for the Internet of Things(n.d.).https://www.ibm.com/blogs/internet-of-things/watson-iot-blockchain/ (accessed Feb 11, 2019).

Why Blockchain Is the Missing Link To IoT Transformations-Maciej Kranz- IDG Contributor Network(n.d.).https://www.networkworld.com/article/3295903/internet-of-things/why-blockchain-is-the-missing-link-to-IoT-transformations.html (accessed Feb 10, 2019).

CHAPTER 14

A Study on Blockchain Technology: Application and Future Trends

GURINDER SINGH[1], VIKAS GARG[1*], and POOJA TIWARI[2]

[1]*Amity University, Uttar Pradesh, India*

[2]*ABES Engineering College, Ghaziabad, Uttar Pradesh, India*

Corresponding author. E-mail: vgarg@gn.amity.edu

ABSTRACT

In the current competitive scenario, everyday new technology is coming. With the latest advancement in technology, blockchain technology has emerged. It has many advantages and applications in various areas such as cryptocurrencies, banking and financial services, many internet things, managing risk etc. There are many researches that were conducted in the concerned area, but there is no research that has conducted to understand the application of this technology and also summarizing the future trends. This article is focusing on the application of blockchain technology in different areas and also understanding the future opportunities that will be opened in different areas due to this technology.

As it says blockchain is made up of two different words: block and chain. Here, block is referred to as the small units of individual transactions done in the cryptocurrency via many of computers across the globe. While chain is referred to as the linkage of all the network computers that combines all the transactions and forms a ledger of all the records. So, from here we come to know blockchain is the record of all the small units of transactions done on the cryptocurrency like Bitcoin, just like a ledger, on a particular network that is connected with all the computers performing these transactions. Blockchain technology is emerging technology just like artificial intelligence (AI). As AI is creating a huge difference in the current scenarios and in different sectors, likewise, blockchain is creating impacts and getting different sectors revaluated.

In this chapter, we are not talking about the whole situation or sector we can say on which blockchain is creating its impact. Rather, the main focus is on the IT sector in which the blockchain technology is creating a huge revaluation on. So, here we discuss the following revaluation that blockchain creates on IT.

14.1   INTRODUCTION

Blockchain technology is digital in nature and its main feature is the recording of the information and data by using a logbook approach having the different essential features: ordered, increased, sound, and digital. In addition to these characteristics, other features of the blockchain are not actually its features but they are due to the addition of certain features such as sharing, distribution, and communication and agreement protocols. Technology of blockchain is a collection of different chains of data linked to each other cryptographically. These blockchains are linked to each other in a sequential manner by cryptographic hashes. Hash is a fixed length size, which is derived from a given document or a message. Three major components were involved in each block are: (1) Block-Data: a set of messages or transactions; (2) Chaining-Hash: a copy of the hash value of the immediately preceding block; and (3) Block-Hash: the calculated hash value of the data block or messages plus the chaining hash value (Norton, 2016; Gupta, 2017; Naka-moto, 2008). The technology of blockchain was proposed in the year 2008, and it was implemented in the year 2009. It is considered as a public ledger, in which blocks are committed toward maintaining the transactions that are stored in a chain of blocks. These blocks in the chain form are continuously growing after the blocks are appended to it. There are certain characteristics of blockchain technology like decentralization, persistency, privacy, and auditability. This technology of blockchain also works under a spread out atmosphere that is permitted by incorporating some fundamental technologies such as cryptographic hash, digital sign, and disseminated consensus technique. Transaction can be done in the decentralized environment by the help of blockchain technology. if the technology of blockchain is adopted, efficiency can be improved. it will also help save the cost (Norton, 2016).

Recently, the attention of both academic and industry is attracted by cryptocurrency. In the year 2016, the financial capital market achieved a great victory with the help of the very initial cryptocurrency, that is, Bitcoin through which it attained the level of $10 billion. Blockchain famous application is Bitcoin, but still this technology can be applied in diverse fields and area beyond cryptocurrencies. Since it facilitates the payment of the bank without the intervention of any intermediary, blockchain can be used in different services of finances like in the assets available digitally, settling, and making digital payments (Foroglou, 2015; Peters, 2015). Additionally, this technology of blockchain is giving attention as the leading technology for the future generations in the field of communicating system of internet like the advance agreements (Kosba, 2016), community services (Akins, 2013), Internet of Things (IoT) (Zhang, 2015), status systems (Sharples, 2015), and safety facilities (Noyes, 2016). But despite of this fact that blockchain technology is emerging as one of the technologies that has a great potential for building the future internet system, but this technology is simultaneously confronting various technical hurdles (Foroglou and Tsilidou, 2016).

Above all, scalability is an enormous worry. The extent of the bitcoin is bound to 1 MB at present and these squares are mined at regular intervals. Consequently, the network of bitcoin is limited to the 7 transactions per second, which is not capable of dealing with high frequency trading. However, the blocks are considered as having a larger space and propagation of the network in a slower way. There is a slow shift to centralization as each and every operator using the blockchain technology wants to conserve the blockchain. Therefore, it is becoming one of the challenges to maintain the tradeoff between size of the block and security. Furthermore, it is quite visible that with the usage of the self-centered excavating plan, the miners are able to attain a huge profit as compared to the true share (Eyal, 2014). Additionally, these blocks are hidden by the miners for generation of the profit in the upcoming time (Carboni, 2014).

Likewise, frequently divisions can happen, which actually delays the growth of the blockchain. Hence, there is a need to provide some solution to fix the problem. Furthermore, it has been proved that even if the transactions are done through public and private keys there is a possibility of privacy leakage (Biryukov, 2014). It is also possible to track the real IP address of the user. Additionally, there are certain serious problems such as proving the work done or proving the status are faced in algorithms of current consensus. For example, too much electricity is wasted in case of proof-of-work, whereas these phenomena that in case of proof-of-stake consensus process, the rich is getting richer. So, in case of development of blockchain technology, it is necessary to address these challenges as quickly as possible. Abundant body of literature on blockchain is available from different sources, such as journal articles, proceedings of different conferences, from different forum posts, blogs, codes, and wikis etc. Tschorsch (2016) has conducted the technical survey regarding the currencies that are decentralized in nature and that also includes Bitcoin (Jaag et al., 2016). As compared to Tschorsch et al., this chapter will be focusing on blockchain technology rather than focusing on the digital currency. Nomura Research Institute has formulated the report regarding blockchain that is technical in nature (NRI, 2015). In contradiction to this study, our study has emphasized on the state of art regarding blockchain researches that also include latest development and future trends.

In other way, blockchain technology helps create a distributed ledger system that is robust, secure, and transparent in nature. This technology is also a revolution in its own way that is not usual and new in nature for maintenance of public databases that can be better understood as the technology that is actually institutional and social technology. This new technology can have an impact on the economy of the country as the approach of public choice/institutional approach to the economics of blockchains (Luther and White, 2016).

14.2   ARCHITECTURE OF BLOCKCHAIN

Blockchain consists of different blocks of sequence, which actually records the entire list of transactions such as public ledger that is conventional in nature. Figure 14.1 given above is representing the example of blockchain. Parent block can be referred to as the block that is previously connected to a block via reference through a hash value. It should be rather important to observe that ethereum blockchain will be able to store the offspring of the ancestral blocks, also known as uncle block hashes.

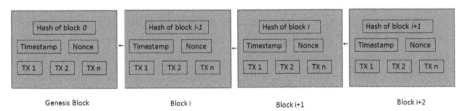

FIGURE 14.1   Architecture of blockchain that consists of different blocks in a sequential manner.

Genesis block is the first block of the sequence of the blockchain, which don't have any parent block. Further in another section of this chapter, we have shown the structure of the blockchain and mechanism of digital signature in another section. Furthermore, we have also discussed the important characteristics and also in further sections certain taxonomy related to blockchain has been discussed (Atzori, 2015).

14.3   BLOCK

A block in blockchain technology includes two things: header of the block and the body of the block as given in the Figure 14.2.

With regard to header of the block, it basically consists of different components such as:

(i)   Block version: It indicates how validation rules follow in set of block.

(ii) Parent block hash: Pointing toward the previous block, this block has value of 256-bit hash.

(iii) Merkle tree root hash: It considers the entire hash value of all the transactions in the block.

(iv) Timestamp: Current timestamp as seconds since 1970-01-01T00:00 UTC.

(v)   nBits: Current hashing target in compact format.

(vi) Nonce: A 4-byte field, which usually starts with 0 and increases for every hash calculation.

Body of the block actually consists of transaction counter and transactions. The size of the block and also the transaction size determine the number of transactions continued by a particular block. To validate authentication of transactions, asymmetric cryptography mechanism is used in the blockchain technology. In the atmosphere where no one can be trusted easily, digital sign grounded on unequal cryptography is helpful. Now, let's take a concise example of the digital sign (Biryukov et al., 2014).

14.4   DIGITAL SIGNATURE

A pair of communal and private keys is owned by each user in blockchain technology. Transactions are signed by using the private keys. In this case,

Block Version	02000000
Parent Block Hash	B6ff0b2b1680a2862a30 ca44d346d9e8910d334b eb48ca00000000000000 00
Merkle Tree Root	9d10aa52ee949386ca93 85695f04ede270dda208 10decd12bc9b048aaab3 1471
Timestamp	24d95a54
nBits	30c31b18
Nonce	Fe9f0864

Transaction Counter

| TX 1 | | TX 2 | | TX n |

FIGURE 14.2   Block structure.

the signature is done using the private key. These digital signatures can be accessed by the public keys and they are spread through the entire network and are visible to everyone. Figure 14.3 shows the illustration of usage of digital sign in the blockchain. This digital sign can be further classified in the two stages: the signing stage and the authentication stage. We can again look at the example of Figure 14.3. In this, when Alice wishes to sign the transaction, initially she has generated the value of hash that was taken from the deal. After which the hash value was encrypted by utilizing the restricted keys and it was sent to a new operator Bob that consisted of original data having the encrypted hash. Bob then verifies the transaction that he has received by comparing it with the decrypted cash and from the received data hash value is taken. The classic computerized signature calculations utilized in Elliptic Curve Digital Signature Algorithm (ECDSA) (Atzori et al., 2010).

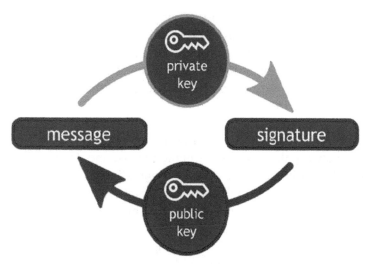

FIGURE 14.3 Usage of digital sign in the blockchain.

14.5 IMPORTANT FEATURES OF BLOCKCHAIN

The characteristics of the blockchain can be categorized as follows:

1. Decentralization: In the centralized transaction system that is conventional in nature, the central trusted agency is responsible for the validation of each transaction (such as the central bank) which in turn certainly results in the budget and the act bottlenecks at the main servers.

TABLE 14.1 Comparisons among Public Blockchain, Consortium Blockchain, and Private Blockchain.

Property	Public blockchain	Consortium blockchain	Private blockchain
Consensus determination	All miners	Selected set of nodes	One organization
Read permission	Public	Could be public or restricted	Could be public or restricted
Immutability	Nearly impossible to tamper	Could be tampered	Could be tampered
Efficiency	Low	High	High
Centralized	No	Partial	Yes
Consensus process	Permissionless	Permissioned	Permissioned

In contrast to this, in the network of the blockchain any two peers, that is, peer-to-peer can perform any transaction without any validation from the main agency. In this approach, it can be observed that the cost of the server can be reduced (comprising the improvement expense and the process expense) and moderate the performance blockages at the main server (Foroglou and Tsilidou, 2015).

- Persistence: As each and every transaction in the blockchain is spread in the network that needs to be authenticated and noted in the blocks disseminated in the entire system, it actually becomes very difficult to tamper. In addition to this, other nodes will be verifying the broadcasted block and each transaction will be checked. So, it becomes easy to detect the falsification.
- Anonymity: In blockchain technology, it is possible for the user to interact by generating the address. Additionally, users can also generate multiple addresses to escape any individuality disclosure. In this, no main person is maintaining the private information of the user. So, in the blockchain this mechanism actually helps maintain the privacy of the transactions up to certain extent. It should be noted that blockchain technology will not guarantee privacy up to the level of perfection due to limitation of internal constraint.
- Auditability: As each transaction in the blockchain technology is recorded and validated by the help of timestamp, so it becomes easy for the user to authenticate and track the former proceedings by opening any transaction in the system circulated (Noyes, 2016).

Each and every transaction can be found which were done formerly in the Bitcoin blockchain technology. This in return helps in increasing the accessibility and transparency in any of the transactions recorded in the blockchain.

14.6   BLOCKCHAIN SYSTEMS TAXONOMY

We can categorize the blockchain system into three different categories: private blockchain, public blockchain, and consortium blockchain. All these different kinds of blockchain system have been compared and shown in Table 14.1.

- Consensus determination: As blockchain is public in nature, so each and every block of it can be used in the process of consensus. But in the consortium blockchain only some of the selected blocks are accountable for the validation of the block. In case of a private chain, in the final consensus it will only be directed by any single association (Peters and Panayi, 2015).

Read authorization: In the unrestricted blockchain, transaction is noticeable in nature whereas the permission of read is dependent on the private blockchain or a group of blockchain. The organization or we can say the consortium has the capacity to choose whether the data that are stored are restricted or public.

- Immutability: As the transactions in the blockchain are kept in several blocks in the system, it becomes very difficult to alter the blockchain that is public in nature. However, if the tampering is to be done majorly at the group or the leading organization, the private blockchain can be reciprocated or altered.
- Effectiveness: There are numerous nodes in the public blockchain network and due to which it consumes a lot of time to broadcast transactions. Considering the safety of the network into consideration, there will be more restriction on the public blockchain. In conclusion to this, the latency will be very high and throughput is limited in the transaction. The efficiency of the consortium and private blockchain can be increased by the fewer validators.
- Centralize: The primary concern of qualification between each of the three sorts of blockchains is that: open blockchain is decentralized, consortium blockchain is in part incorporated and private blockchain is completely brought together as it is constrained by a solitary gathering. (Wanget al., 2015).
- Consensus procedure: Consensus process can be joined by each person in the world in the public blockchain. As compared to unrestricted blockchain, both groups of blockchain are permitted. In the consortium or public blockchain, it becomes important to certify the nodes, which can join the consensus process. Many users can be attracted toward the blockchain as the unrestricted blockchain is exposed to the world. It has been analyzed that unrestricted blockchain has emerged day by day. In case of group blockchain, it can be functional to the various areas of business as well. Presently, Hyperledger has developed the frameworks for the business consortium. Ethereum have also provided the different tools for the formation of the consortium blockchains. In the case of the private blockchains, many companies are using and implementing the

blockchain so as to increase the auditability and efficiency (Zhang and Wen, 2015).

14.7   RECENT ADVANCES AND CHALLENGES IN BLOCKCHAIN

Blockchain technology is on the developing technology, it is confronting numerous threats and difficulties. We have summarized the three basic threats:

1.   Scalability
2.   Privacy leakage
3.   Selfish mining

As every day the numbers of transactions are increasing, it is also influencing the heaviness of the blockchain. Presently, the blockchain of bitcoin has gone beyond 100 GB storage. All these transactions are stored so as to validate each transaction. Additionally, as the restriction was there in the original block size and time gap for the generation of a new block, seven transactions will be processed by the Bitcoin transactions, which in turn will not be able to satisfy the necessity of handling millions of transactions in actual time fashions. In the meantime, as the block's capacity is quite low, it may also delay numerous transactions as miners prefer the transactions with the high transactions fees (Sharples and Domingue, 2015). On the other hand, propagation speed will be slowed down due to the large size of the blockchain and headed to blockchain branches. So, the issue of scalability is relatively challenging and difficult .There is various solutions that are proposed to resolve the challenge of scalability problem that can further be divided into two kinds:

• Optimization of the storage capacity of blockchain: To address the issue of heaviness in the blockchain, a new and innovative scheme of cryptocurrency was proposed. In the new scheme, networks generally remove the transactions that are old and a tree that is referred as database and that is utilized to keep the balance of each and every filled account holders. In the similar method, all the transactions should not be stored by the nodes to verify the validity of the transactions. Additionally, the problems can also be fixed by the lightweight of the client. A novel scheme named VerSum has actually provided a different option by showing the existence of lightweight clients.

VerSum has allowed outsourcing the extensive computation of larger inputs through lightweight clients. It is ensured that the results are compared with the multiple servers and computing the results.

- Redesigning blockchain: Bitcoin-NG was proposed as an emergence of the blockchain technology. The basic propaganda of Bitcoin-NG is to differentiate the conservational blocks into two types: one as the kep block that will be used for electing leaders and the other one as microblock that will be used to keep the transactions. There is competition between the miners to become a leader. It has been analyzed that the new leader will appear only when microblock generation will be generated (Namecoin, 2014).

14.8   FUTURE TRENDS

The blockchains can be implemented in various ways, particularly in the fields that traditionally depend on the third parties to build an assured sum of trust. Atzori (2015) recommends that the blockchain might rebuild the entire society as well as the politics. With the usage of the decentralized stages by the people for the organization and protection of the society, many roles might become outdated. Further, he ends with that self-controlling of the services offered by the government via allowed blockchains is liked and wanted, as it can remarkably improve the functioning of the public management. "decentralization of taxpayer supported organizations through permissioned blockchains is conceivable and attractive, since it can essentially expand open organization usefulness." In the poor nations, the main important task is to restructure the society. With the usage of the blockchain, the wealth of the people is more effectively secured. Particularly in the third world, if for instance the local government targets the population to acquire, then the landlords face the issue of proving their ownership. By indulging these land titles in the blockchain, these existing obstacles can be minimized. Glaser (2017) suggested that the border among the digital empire and the real world could result in the frail connections that harm the digital belief built by the blockchain system. According to the Federal Bureau of Investigation 2012, there is an argument going on among the investigators and controllers that if the cryptocurrency depending on the blockchain can satisfy the purposes of the actual money. Mishkin (2004) has defined money as anything that is commonly recognized for buying goods and services or for the settlement of the debtors. "Anything that is commonly acknowledged in installment for merchandise or benefits or in the reimbursement of obligations." According to Luther and White (2014), cryptocurrencies are hardly used as the mode of give-and-take in today's world. Glaser (2014) provides practical visions that Bitcoin is mainly used as a hypothetical asset. Establishment of the cryptocurrencies as a replacement of old money might make it easier for spending and accepting because of the advanced methods developed by the entrepreneurs. There might be a great contribution of the blockchain in the method of payment done by people in the physical world. While purchasing property, land owners pay huge transaction costs. Goldman Sachs (2016) suggests that for the reduction in the faults and physical effort blockchain could be used so as to decrease the premium of the insurance and could produce $2–4 billion by saving the cost in the US. "Blockchain could reduce title insurance premiums and generate $2–4 billion in cost savings in the US by reducing errors and manual effort." On one hand, the primary focus of the computer scientists is on the technical and cryptographic obstacles in this field, whereas on the other hand, the investigators from the business and information systems engineering profile get the chance to emphasis on the marketplace design, issues of faith and confidentiality, and whether to adopt or not to adopt the innovative technology. Besides, this disruptive advancement might modify various on-going business models and generate innovative ones, and might have huge influence on the whole industries. Hence, it is important to conduct study at the juncture of markets, technology, and business models.

The main four trends of blockchain technology:

There are varieties of trends in blockchain technology. In this segment, we precise various types of trends of blockchain. The authors have classified the trends into different sections in which the first section carries finance; the next one carries IoT. A further section carries public and social services, which is followed by reputation system; and the last section carries security and privacy.

Now, let's take a look into each of the sections.

Finance

1. Traditional business and financial services are largely affected by the evolution of the blockchain systems like Bitcoin (Nakamoto, 2008). Peters, 2015 suggested that the blockchain has the power to disturb the banking world. There are several ways in which blockchain technology can be used such as for the clarification and settlement of the financial assets etc. In addition, Morini (2016) discussed that

there are some of the actual business cases like collateralization of the financial derivatives in which usage of blockchain can decrease the charges and threats. Many software companies like Microsoft Azure (2016) and IBM (2016) are planning to introduce blockchain as a service as it is seeking great attention from these companies.

2. Blockchain also helps the old-style organizations to easily complete the transformation of the enterprise likewise the advancement of the business and financial services. Let's take an example of the postal operators (POs) as the basic role of postal operators is to perform as an intermediary among the customers and the merchants, whereas blockchain and cryptocurrency help in extending the basic role of the postal operators through the advancement of the innovative financial as well as non-financial services. Jaag and Bach (2016) discovered that a kind of colored Bitcoin known as postcoin can be issued by each and every postal operator on their own, which is a great chance of advancement for the postal operators through the blockchain technology. With the huge retail network of postal operators, postcoin can be easily availed, as the postal operators are seen as the most trustworthy authority by the people. It is also discussed in Jaag and Bach (2016) that in addition to this, blockchain technology also provides other opportunities of business to the postal operators such as identification of the services, management of the devices, and management of the supply chain.

3. Also known as P2P, so this blockchain technology helps in building a more safe and dependable P2P financial market. Noyes find out many methods of merging multiparty calculation protocols with the P2P tools so as to come up with a new Peer-to-Peer Financial Multiparty Calculation market (Noyes, 2016). Blockchain-based MPC showcase permits offloading computational undertakings onto a system of mysterious companion processors.

4. The structure of the management of risk plays an important role in the financial technology (also known as FinTech) and with its combination with the blockchain technology its performance can be advanced. Pilkington (2016) came up with a new outline of risk management, in that blockchain technology is utilized for the analysis of the risk involved in the investment in the Luxembourgish situation. Nowadays, investors who have the securities through the links of defenders have the high chances of facing risk of the failure in any of the securities. Blockchain technology helps in deciding the

investments and securities at a faster rate rather than going through the long-term process. Micheler (2016) shows that a combination of a new system with the blockchain technology can decrease the custodial risk by maintaining the equal level of safety in the transactions. Other than this, blockchain also empowers the self-ruling organizations to participate in the collaborations related to the business work. A profoundly trustworthy DAO-GaaS conflict model (Norta 2015) was proposed to defend business-semantics incited consistency rules.

Internet of Things

One of the technologies that are gaining the attention is IoT, that is, Internet of Things, which is arising as the peak favorable technology for information and communication (ICT). The purpose of IoT is to deliver its users with different kinds of services by incorporating the things (also known as smart objects) into the internet (Atzori, 2010; Miorandi, 2012). The basic important applications of Internet of Things incorporated for management of logistic with Radio Frequency Identification (RFID) technology (ISO/IEC, 2013), savvy homes (Dixon, 2012), e-Health (Habib, 2015), savvy grids (Fan, 2013), maritime Industry (Wang, 2015), and so on.

Hypothetically, blockchain technology can bring a lot of improvement in IoT.

1. Zhang (2015) suggested an innovative e-business model built on blockchain technology with the understanding of business deal of savvy property and savvy agreement. According to this model, the self-ruling transaction entity is being assumed by the Distributed Autonomous Corporations (DAC). People transact with DAC to acquire coins and interchange the sensor facts deprived of any intermediary (Dixonet al., 2015).
2. Protection of security and confidentiality is one of the crucial concerns of the IoT industry. Privacy of the IoT applications can be improvised with the help of the blockchain. Hardjono (2016) specifically suggested a privacy–security technique for assigning into the cloud environment through an IoT device. More precisely, an innovative construction was suggested in Hardjono (2016) to support the device to demonstrate its producing source deprived of the verification of the intermediary and it is permitted to list unknowingly. Other than this, IBM (2015) disclosed its evidence of notion for Autonomous Decentralized Peer-to-Peer Telemetry (ADEPT), which is a structure

that uses blockchain technology to construct a circulated set-up of devices. In this model (ADEPT), machines working in the household will be capable of identifying the operational issues and recover the software updates itself.

Community and Communal Services

In community and communal services, there is a broad use of blockchain technology (Carboni, 2016).

1. In public services, the basic blockchain application is registering the land (NRI, 2015). in this, blockchain can be used to register and broadcast the information regarding the land like the actual status of the land and the related rights of it. Other than this, if any modifications are done on the land status like transfer of land or formation of any kind of mortgage can be documented and recorded on the blockchains that will directly increase the proficiency of public services.

2. Further, green energy can also make use of blockchains. Gogerty (2011) suggested solar coin to boost the utilization of renewable energies. In this, solar coins are a type of digital currency that is satisfying the solar energy makers. Other than the traditional way of attaining the coins via mining, these solar coins can be arranged by the solar coin foundation till the time span you create the solar energy.

3. In the trust less atmosphere, blockchain is formerly invented to carry out the transactions of the currency. Though the technique of blockchain can also be useful to the digital era of the Education industry, if we continue with the knowledge and training procedure as the currency. In Devine (2015), learning of blockchain was suggested. In this learning of blockchain, the blocks can be assumed packed and positioned into the blockchain by the academicians, and the coins can be assumed as the learning accomplishments (Hardjono and Smith, 2016).

4. Additionally, the internet set-up like DNS and identities can be protected by the blockchain. For instance, Namecoin (2014) is a new open-source technology that is used to increase safety, decentralization, censorship conflict, confidentiality, and the speed of the DNS and identities (Namecoin, 2014). By creating the web stronger to censorship, it guards the online free-speech rights. According to Akins (2013), blockchains can be further utilized to other public services like registration of marriage, management of patents, and system of income taxation. With the involvement of the blockchain

in the public services, replacement of seals attached on the documents might be the mobile devices with digital signatures entrenched in them, which are given to the administrative branches. With the help of this, there will be significant saving in the huge paperwork (Goldman Sachs, 2016).

Reputation System

Reputation is one of the significant factors that show the ample of faith, belief, and trust in the community regarding you. There is a direct relation between the reputation and your trustworthiness among others, the higher the reputation, the greater the trustworthiness is there for you by others. The prior transactions and communications with the public act as a parameter on which the evaluation of the reputation of any person can be done. There is an increase in the number of cases in which fake personal reputation is shown. For instance, in the e-commerce industry, numerous service providers register a large number of fraud clients to attain a great status. For solving this issue, blockchain is the solution.

1. In academics, reputation plays a vital role. Sharples (2015) suggested there should be involvement of the blockchain-based dispersed set-up for the reputation and scholastic records. In the initial step, each and every institution and knowledgeable employee will be provided with a preliminary award scholastic reputation currency. Any institution can honor an employee by conveying certain of its reputation accounts to the employees. As all the transactions are recorded on the blockchain, then it is simple to notice any change in the reputation.

2. The skill to measure the status of any associate of the net community is very significant. Carboni (2015) suggested a blockchain-based model of reputation, in which any customer pleased with the facility and is willing to provide a positive response, and then a voucher will be duly signed by them. For the discouragement of the Sybil attack, the service provider will be liable to pay additional 3% of the expense to the system as the elective fee, after getting the signature of the customer on the voucher. Reputation of the service provider can be computed on the basis of the total of the voting fee. Dennis (2015) suggested an innovative set-up of reputation that is specifically applicable for the diverse networks. In specification, they produced an innovative blockchain that can

record any one-dimensional value of reputation such as 0 or 1 from the finalized transactions. For instance, A sends any file to B. after receiving the file, B further sends a transaction that includes the score, hash of file, and the private key of B for the verification of the identity. After this, the miners will contact both A and B so as to confirm that the transaction is occurring without any suspicion. It is impossible to manipulate the records of the reputation as all the transactions are recorded on the blockchain (Akins et al., 2015).

14.9 SECURITY AND PRIVACY

Safety Improvement: Due to the increased usage of different cell phones and mobile services, which are additionally showing their weakness to noxious hubs? There are numerous anti-malware strainers suggested to find out the suspicious records via some identical pattern systems, which are the main server to keep and modify the bug. Though these centralized defense actions are also defenseless to some mischievous attackers. The security of widespread networks can be improvised with the help of the blockchain. In specific, Noyes (2016) suggested a new anti-malware atmosphere that is known as BitAV, in which the user can allot the pattern of the virus on the blockchain. With this technique, BitAV can improve the error patience. As shown in Noyes (2016), BitAV can be used to increase the speed of scanning as well as to improve the fault dependability, that is, decrease in the vulnerability to directed ignorance of the service attacks. Dependability of the security set-up can also be improvised with the help of blockchain technology. For instance, conservative Public Key Infrastructure (also known as PKIs) is frequently suspected to one point of failure because of the faults in the hardware and the software or mischievous attacks. In Axon (2015), it is shown that blockchain can also be utilized in building privacy conscious PKI while hand-in-hand refining the dependability of the conservative PKIs.

Privacy Protection: With the increasing risk of disclosure of the personal information to the malwares, many cell phone services and public web sources are gathering our sensitive data. For instance, Facebook has gathered more than 300 petabytes of private information since the day of its commencement (Vagata, 2014). Typically, the gathered facts are kept at the main servers of the service providers, which are suspected to be the mischievous outbreaks.

For the improvement of the safety of the private delicate information, block-chain can be used. In suggestion a self-ruling private information managing set-up that safeguards the user possession of their personal information (Gupta, 2014). This set-up is applied on blockchain. The privacy problems in which this system can safeguard the data are as follows:

- Ownership of data
- Auditability and transparency of data
- Control on fine-grained accessibility of data

Alike this system, one set-up relied on blockchain technology was also suggested to safely allocate delicate information in a self-ruling way in Ethos (2014).

KEYWORDS

- **Bitcoin**
- **artificial intelligence**
- **YUCA world**
- **hacking**
- **cryptocurrency**

REFERENCES

Akins, B. W.; Chapman, J. L.; Gordon, J. M.: A Whole New World: Income Taxconsiderations of The Bitcoin Economy. 2013, https://ssrn.com/abstract=2394738.

Atzori, M (2015) Blockchain Technology and Decentralized Governance: Is the State Still Necessary? Work Pap Barber S, Boyen X, Shi E, Uzun E (2012) Bitter to Better—How to Make Bitcoin a Better Currency. International Conference on Financial Cryptography and Data Security. Springer, Heidelberg, pp 399–414.

Atzori, L.; Iera, A.; Morabito, G. The Internet of Things: A Survey. (15), 2787–2805.

Axon, L.. Privacy-Awareness in Blockchain-based PKI. CDT Technical Paper Series, 2015.

Biryukov, A.; Khovratovich, D.; Pustogarov, I. Deanonymisation of Clients in Bitcoin p2p Network. In: Proceedings of the 2014 ACM SIGSAC Conference on Computer and Communications Security. pp. 15–29. New York, NY, USA, 2014.

Carboni, D.(2015). Feedback Based Reputation on Top of the Bitcoin Blockchain. arXiv preprint arXiv:1502.01504.

Devine, P. Blockchain Learning: Can Crypto-currency Methods be Appropriated to Enhance Online Learning? In: ALT Online Winter Conference, 2015.

Dixon, C.; Mahajan, R.; Agarwal, S.; Brush, A.; Saroiu, B. L. S.; Bahl, P. An operating system for the home. In: NSDI. USENIX, 2012.

Ethos (2014), http://viral.media.mit.edu/projects/ethos/

Eyal, I.; Sirer, E. G. Majority is not enough: Bitcoin mining is vulnerable. In: Proceedings of International Conference on Financial Cryptography and Data Security. pp. 436–454. Berlin, Heidelberg, 2014.

Fan, Z.; Kulkarni, P.; Gormus, S.; Efthymiou, C.; Kalogridis, G.; Sooriyabandara, M.; Zhu, Z.; Lambotharan, S.; Chin, W. H. Smart Grid Communications: Overview Of Research Challenges, Solutions, And Standardization Activities. (1), 21–38.

Federal Bureau of Investigation (2012) Bitcoin Virtual Currency: Intelligence Unique Features Present Distinct Challenges for Deterring Illicit Activity. https://www.wired.com/images_blogs/threatlevel/2012/05/Bitcoin-FBI.pdf. Accessed 30 Nov 2016.

Foroglou, G.; Tsilidou, A. L. Further Applications of the Blockchain, 2015.

Glaser, F. Pervasive Decentralisation of Digital Infrastructures: A Framework for Blockchain Enabled System and Use Case Analysis, 2017.

Glaser, F., Zimmermann, K.: Haferkorn, M.; Weber, M.; Siering, M. Bitcoin-asset or Currency? Revealing Users' Hidden Intentions. In: Proceedings of the 22nd European Conference on Information Systems (ECIS 2014); Tel Aviv, Israel, 2014.

Gogerty, N.; Zitoli, J. Deko: An Electricity-backed Currency Proposal. Social Science Research Network, 2011.

Goldman Sachs Profiles in Innovation – Blockchain. http://www.the-blockchain.com/docs/Goldman-Sachs-report-Blockchain-Putting-Theory-into-Practice.pdf. Accessed 30 Nov 2016.

Gupta, M. Blockchain for Dummies, IBM Limited Edition, John Wiley and Sons, Hoboken, NJ, available at: https://public.dhe.ibm.com/common/ssi/ecm/xi/en/xim12354usen/XIM12354USEN.PDF

Habib, K.; Torjusen, A.; Leister, W. Security Analysis of a Patient Monitoring System for the Internet of Things in eHealth. In: The Seventh International Conference on eHealth, Telemedicine, and Social Medicine (eTELEMED), 2015.

Hardjono, T.; Smith, N. Cloud-based Commissioning of Constrained Devices Using Permissioned Blockchains. In: Proceedings of the 2nd ACM International Workshop on IoT Privacy, Trust, and Security. pp. 29–36. ACM, 2016.

IBM Blockchain.http://www.ibm.com/blockchain/the blockchain. In: 2015 10th International Conference for Internet Technology and Secured Transactions (ICITST). pp. 131–138. IEEE,2015.

IBM: IBM ADEPT Practitioner Perspective - Pre Publication Draft, 2015.

ISO/IEC 18000 (2013), http://en.wikipedia.org/wiki/ISO/IEC_18000

Jaag, C.; Bach, C.; et al. (2016). Blockchain technology and cryptocurrencies: Opportunities for postal financial services. Tech. rep.

Kosba, A.; Miller, A.; Shi, E.; Wen, Z.; Papamanthou, C. Hawk: The Blockchain Model of Cryptography and Privacy-Preserving Smart Contracts. In: Proceedings of IEEE Symposium on Security and Privacy (SP). pp. 839–858. San Jose, CA, USA, 2016. Luther, W. J.; White, L. H. Can Bitcoin Become a Major Currency? Working Paper. Miers, I, Garman, C, Green, M, Rubin, A. D. (2013) Zerocoin: Anonymous Distributed e-cash from Bitcoin. IEEE Symposium on Security and Privac. IEEE. pp 397–411, 2014.

M. Nofer et al. Blockchain, Bus Inf Syst Eng 59(3):183–187 (2017) In: Proceedings of the 50th Hawaii International Conference on System Sciences (HICSS 2017), Waikoloa Village, Hawaii, 2017.

Micheler, E.; von der Heyde, L. Holding, Clearing And Settling Securities Through Blockchain Technology Creating an Efficient System by Empowering Asset Owners. Social Science Research Network.

Microsoft Azure: Blockchain as a Service. (2016), https://azure.microsoft.com/en-us/solutions/blockchain/

Miorandi, D.; Sicari, S.; Pellegrini, F. D.; Chlamtac, I. Internet of Things: Vision, Applications and Research Challenges.   (7), 1497–1516.

Mishkin, F. S. The Economics of Money and Financial Markets, 7th Edn. Pearson: Boston, 2004.

Morini, M. From 'Blockchain Hype' to a Real Business Case for Financial Markets. Social Science Research Network, 2016.

Nakamoto, S. Bitcoin: A Peer-To-Peer Electronic Cash System (2008), https://bitcoin.org/bitcoin.pdf

Namecoin (2014), https://www.namecoin.org/

Norta, A.; Othman, A. B.; Taveter, K. Conflict-resolution Lifecycles for Governed Decentralized Autonomous Organization Collaboration. In: Proceedings of the 2015 2nd International Conference on Electronic Governance and Open Society: Challenges in Eurasia. pp. 244–257. ACM, 2015.

Norton, S. (2016), "CIO explainer: what is blockchain?", The Wall Street Journal Blog, available at: https://blogs.wsj.com/cio/2016/02/02/cio-explainer-what-is-blockchain/

Noyes, C. (2016). Bitav: Fast Anti-Malware by Distributed Blockchain Consensus and Feedforward Scanning. arXiv preprint arXiv:1601.01405

Noyes, C.: Efficient Blockchain-Driven Multiparty Computation Markets at Scale. Tech. rep. (2016).https://www.overleaf.com/articles/blockchain-multiparty-computation-markets-at-scale/mwjgmsyybxvw/viewer.pdf

NRI: Survey on Blockchain Technologies and Related Services. Tech. rep., 2015.

Peters, G. W.; Panayi, E. Understanding Modern Banking Ledgers Through Blockchain Technologies: Future of Transaction Processing And Smart Contracts on the Internet of Money. Social Science Research Network, 2015.

Peters, G. W.; Panayi, E.; Chapelle, A. Trends in Crypto-Currencies and Blockchain Technologies: A Monetary Theory and Regulation Perspective (2015), http://dx.doi.org/10.2139/ssrn.2646618

Pilkington, M. Blockchain Technology: Principles and Applications. In , Olleros, F. X., Zhegu, M., Eds. Edward Elgar. Available at SSRN: http://ssrn.com/abstract=2662660

Richard Dennis, G. O. In , Proceedings of the 10th International Conference for Internet Technology and secured Transactions , 2015, pp. 131–138.

Sharples, M.; Domingue, J. The Blockchain and Kudos: A Distributed System for Educational Record, Reputation and Reward. In: Proceedings of 11th European Conference on Technology Enhanced Learning (EC-TEL 2015). pp. 490–496. Lyon, France, 2015.

Tschorsch, F., Scheuermann, B. Bitcoin and Beyond: A Technical Survey on Decentralized Digital Currencies. 18(3), 2084–2123.

Vagata, P.; Wilfong, K. Scaling the Facebook Data Warehouse to 300 PB. Tech. rep. (2014),   https://code.facebook.com/posts/229861827208629/scaling-the-facebook-data-warehouse-to-300-pb/

Wang, H.; Osen, O.; Li, G.; Li, W.; Dai, H. N.; Zeng, W. Big Data and Industrial Internet Of Things For The Maritime Industry In Northwestern Norway. In: IEEE Region 10 Conference (TENCON), 2015.

Zhang, Y.; Wen, J. An IoT Electric Business Model Based on the Protocol of Bitcoin. In: Proceedings of 18th International Conference on Intelligence in Next Generation Networks (ICIN). pp. 184–191. Paris, France, 2015.

Index